T0222415

The Cognitive Structure of Scientific Revolutions

Thomas Kuhn's *Structure of Scientific Revolutions* became the most widely read book about science in the twentieth century. His terms "paradigm" and "scientific revolution" entered everyday speech, but they remain controversial. In the second half of the twentieth century, the new field of cognitive science combined empirical psychology, computer science, and neuroscience. In this book, recent theories of concepts developed by cognitive scientists are used to evaluate and extend Kuhn's most influential ideas. Based on case studies of the Copernican revolution, the discovery of nuclear fission, and an elaboration of Kuhn's famous "ducks and geese" example of concept learning, the volume offers new accounts of the nature of normal and revolutionary science, the function of anomalies, and the nature of incommensurability. This new approach to the intellectual content of science and its historical development incorporates insights from both traditional philosophy of science and constructivist sociology of science. The main technique presented, the dynamic frame model of human concepts, may be applied to any field where the nature of concepts is important.

Hanne Andersen is associate professor of history and philosophy of science at the University of Aarhus, Denmark.

Peter Barker is professor of history of science at the University of Oklahoma.

Xiang Chen is professor of philosophy at California Lutheran University.

The Cognitive Structure of Scientific Revolutions

HANNE ANDERSEN

University of Aarhus

PETER BARKER

University of Oklahoma

XIANG CHEN

California Lutheran University

CAMBRIDGE
UNIVERSITY PRESS

CAMBRIDGE UNIVERSITY PRESS
Cambridge, New York, Melbourne, Madrid, Cape Town,
Singapore, São Paulo, Delhi, Mexico City

Cambridge University Press
The Edinburgh Building, Cambridge CB2 8RU, UK

Published in the United States of America by Cambridge University Press, New York

www.cambridge.org
Information on this title: www.cambridge.org/9781107637238

First published 2006
First paperback edition 2013

A catalogue record for this publication is available from the British Library

Library of Congress Cataloguing in Publication Data
Andersen, Hanne, 1964–
The cognitive structure of scientific revolutions / Hanne Andersen, Peter Barker,
Xiang Chen.
p. cm.
Includes bibliographical references and index.
ISBN-13: 978-0-521-85575-4
ISBN-10: 0-521-85575-6
1. Science – Philosophy – History – 20th century. 2. Science – History –
20th century. 3. Paradigm (Theory of knowledge) 4. Cognition.
5. Constructivism (Philosophy) 6. Kuhn, Thomas S. I. Barker, Peter, 1949– .
II. Chen, Xiang, 1954– . III. Title.
Q174.8.B365 2005
509'.04–dc22   2005021713

ISBN 978-1-107-63723-8 Paperback

For
C.S.
C.M.W.
and
L.L.

Perhaps the best way to express our position is by proposing a ten year moratorium on cognitive explanations of science. . . . We hereby promise that if anything remains to be explained at the end of this period, we too will turn to the mind!

Bruno Latour and Steve Woolgar, 1986

Contents

Figures

Acknowledgments

Over the last ten years, the three authors of this book have collaborated on a series of studies applying ideas from cognitive psychology to issues in the philosophy of science, and particularly to the work of Thomas Kuhn. In our previous papers we have included a statement that each work was a collaboration to which all three authors contributed equally. The same is true of the current book, which consolidates and extends our ten years of joint work. None of us could have written this book without the help of the others; our discussions have now continued for so long that it is not appropriate to connect particular points in the overall argument with individual contributors.

Our earlier papers on the themes treated in this book include, in chronological order: Andersen, H., Barker, P., and Chen, X. (1996), "Kuhn's mature philosophy of science and cognitive science," *Philosophical Psychology*, 9: 347–363, used by permission of the Taylor & Francis Group (http://www.tandf.co.uk); Andersen, H. (1996), "Categorization, anomalies, and the discovery of nuclear fission," *Studies in History and Philosophy of Modern Physics* 27: 463–492, © 1996, material used here by permission of Elsevier; Chen, X., Andersen, H., and Barker, P. (1998), "Kuhn's theory of scientific revolutions and cognitive psychology," *Philosophical Psychology* 11: 5–28, used by permission of the Taylor & Francis Group (http://www.tandf.co.uk); Andersen, H. (2000), "Kuhn's account of family resemblance: A solution to the problem of wide-open texture," *Erkenntnis* 53: 313–337, © 2000, with kind permission of Springer Science and Business Media;

Andersen, H. (2000), "Learning by ostension: Thomas Kuhn on science education," *Science & Education* 9: 91–106, © 2000, with kind permission of Springer Science and Business Media; Chen, X., and Barker, P. (2000), "Continuity through revolutions: A frame-based account of conceptual change," *Philosophy of Science (Proceedings)* 67: 208–223, © 2000 by the Philosophy of Science Association, all rights reserved; Andersen, H. (2001), "Reference and resemblance," *Philosophy of Science (Proceedings)* 68: S50–S61, © 2001 by the Philosophy of Science Association, all rights reserved; Barker, P. (2001), "Kuhn, incommensurability and cognitive science," *Perspectives on Science* 9: 433–462, © the Massachusetts Institute of Technology; Barker, P. (2001), "Incommensurability and conceptual change during the Copernican Revolution," in P. Hoyningen-Huene & H. Sankey (eds.), *Incommensurability and Related Matters*, Boston Studies in the Philosophy of Science (Boston: Kluwer), 241–273, material used with kind permission of Springer Science and Business Media; Chen, X. (2002), "The 'platforms' for comparing incommensurable taxonomies: A cognitive-historical analysis," *Journal of General Philosophy of Science* 33: 1–22, © 2002, used with kind permission of Springer Science and Business Media; Barker, P., Chen, X., and Andersen, H. (2003), "Kuhn on concepts and categorization," in T. Nickles (ed.), *Thomas Kuhn* (Cambridge: Cambridge University Press), 212–245, material reprinted with permission.

While portions of the present work recapitulate or rework material presented in some of these papers, this book presents a new setting for all our earlier work. In addition to new historical material, we here present completely new accounts of the nature of anomaly, the nature of normal and revolutionary science, and the nature of incommensurability, which supersede the accounts given in our previous papers. This book is the only complete statement of our current views.

The authors wish to thank Adena Alvis, Roger Ariew, Lawrence W. Barsalou, William Bechtel, William F. Brewer, Bernard R. Goldstein, Catherine Hobbs, Li Ping, Nancy Nersessian, Stig Andur Pedersen, and Sylwester Ratowt for their support and advice at different times during the writing of this book, without necessarily implying that they endorse the opinions expressed here.

Hanne Andersen acknowledges the Danish Institute for Advanced Studies in the Humanities (Danmarks humanistiske forskningscenter), the Carlsberg Foundation, and the Danish Natural Science

Research Council, for supporting different portions of the research contributing to this work.

Peter Barker acknowledges the support of a sabbatical leave from the University of Oklahoma and a senior research fellowship from the Danish Institute for Advanced Studies in the Humanities (Danmarks humanistiske forskningscenter), together with assistance from the University of Copenhagen and Danmarks Nationalbank.

Xiang Chen acknowledges the support of Hewlett grants from California Lutheran University.

The Cognitive Structure of Scientific Revolutions

1

Revolutions in Science and Science Studies

1.1 THE PLACE OF KUHN'S WORK IN STUDIES OF SCIENCE

Thomas Kuhn's *Structure of Scientific Revolutions* became one of the most influential books of the twentieth century, although its author suffered the fate of many prophets: he was ignored by the people he most hoped to influence. His technical terms became so widely known that a popular cartoonist could depict a newly hatched chick greeting the world with the cry "Oh! Wow! Paradigm shift!" (Taves 1998) and a best-selling guide to success in life and business would tell its readers, "[W]e need to understand our own 'paradigms' and how to make a 'paradigm shift'" (Covey 1990: 26). But there is no Kuhnian school of history, and many philosophers of science remain skeptical about his ideas. At the close of the twentieth century philosophers generally rejected paradigm shifts and normal science as useful categories for understanding scientific change and were still arguing about another key idea, incommensurability (Curd and Cover 1998; Hoyningen-Huene and Sankey 2001). Meanwhile Kuhn's emphasis on the historical variability of scientific standards and the role of research communities in scientific change was embraced by a new generation of sociologists of scientific knowledge. The new sociologists of science adopted Kuhn as a founding father, if not an intellectual guide: Kuhn's emphasis on the cognitive content of science was marginalized. Our aim in this book is to rectify this situation, by legitimizing the study of the cognitive content of science, in a new way, and providing the tools needed to write a defensible cognitive history of science. At the same time we hope to restore the ideas of conceptual revolutions and incommensurability to the central position they deserve in academic and practical studies of science.

Kuhn's notion of incommensurability provoked especially intense criticism from philosophers, who rejected his early account and largely ignored later attempts to dispel misunderstandings and refine or vindicate the notion through detailed studies of conceptual change in science (Hoyningen-Huene 1993; Kuhn 2000). There were many reasons for this; one of the most weighty was the conflict between mainstream English language philosophy and the theories of concepts developed by Kuhn and other cognitively inclined philosophers of science as the foundation for their work on scientific change. At the same time that Kuhn was refining his theory of concepts, empirical research in cognitive psychology and cognitive science began to undermine the classical theory of concepts, thus providing a new kind of support for Kuhn's philosophical account of science, and especially his account of scientific change. In this book we will use techniques from cognitive psychology and cognitive science to support and extend Kuhn's ideas on the nature of science. Our aim is to recover insights about revolutions and incommensurability in a form that will be usable by philosophers, historians, sociologists, and others who study science and its history.

1.2 REVOLUTIONS IN SCIENCE

Throughout this book we shall draw on detailed case studies of very different developments in the history of science. Two we will present in considerable detail, and two more briefly. We will present detailed examinations of the Copernican revolution, from the midsixteenth to the early seventeenth century, and of the discovery of nuclear fission during the third decade of the twentieth century. While the former has long been discussed as a key episode in the origins of modern science, the latter had equally important consequences inside and outside science. We will supplement these historical case studies with a briefer examination of developments in nineteenth-century ornithology, when the introduction of Darwin's theory led to changes in the classification of birds. We shall argue that in all of these cases, the conceptual structures develop in ways that display several revolutionary traits.

The discovery of nuclear fission was clearly a revolutionary development. In December 1938, Otto Hahn and Fritz Strassmann in Berlin performed an experiment with uranium that had unexpected results. They seemed to have created barium, an element with a nucleus scarcely half the size of uranium. Hahn and Strassmann asked the Austrian exile Lise Meitner for help, and assisted by her nephew, Otto Frisch, she explained how this strange thing could happen. Meitner and Frisch proposed that when struck by a neutron, the atomic nucleus was capable of disintegrating into two roughly equal fragments, releasing a great deal of energy and several additional neutrons. The practical implications of this discovery are well known (e.g., Flügge 1939). As word of Meitner and Frisch's interpretation spread, the international community of physicists rapidly accepted a new idea at radical variance with conventional wisdom.

The general acceptance of Meitner and Frisch's interpretation of the Berlin experiments also called into question an entire class of previously accepted research results that had seemed to establish the existence of a whole class of transuranic elements. These 'discoveries' had been made by Fermi's research group, and others, in earlier neutron bombardment experiments. After the general acceptance of Meitner and Frisch's proposal, all such experiments had to be reevaluated. In the opening stages of the Second World War, the previous results on transuranic elements were retracted, and the discovery of transuranics was recertified, on the basis of the work of Seaborg and Segrè, between 1939 and 1942 (Seaborg 1989).

The nature of the change that occurred in science in 1939 contrasts surprisingly with the events surrounding the supposed discovery of transuranic elements earlier in the decade. The technique of neutron bombardment had become available only after the discovery of the neutron in 1932. The use of a new technique to create completely new elements – elements not found in nature – might well have been expected to cause controversy. However, the Fermi group's claim to have created transuranic elements by neutron bombardment of uranium was accepted rapidly and without any major dislocations elsewhere in the structure of scientific knowledge.

To complicate matters further, the possibility that the nucleus could split into two relatively equal fragments had been suggested by a German scientist, Ida Noddack, in 1934, four years prior to the discovery of nuclear fission. Noddack suggested that "[w]hen heavy nuclei are bombarded by neutrons, it is conceivable that the nucleus breaks up into several large fragments, which would of course be isotopes of known elements" (Noddack 1934b). This suggestion was ignored or dismissed by the same community that rapidly accepted Meitner and Frisch's interpretation of the phenomena in 1939. Cognitive analysis can explain why the discovery of transuranic elements scarcely created a ripple on the surface of science, why the discovery of fission had so much more profound effects, and why the same community that rejected fission in 1934 accepted it in 1939.

Kuhn did not examine the discovery of fission in *The Structure of Scientific Revolutions*, although he did consider a wide range of the historical cases, most prominently the transition from the phlogiston theory to Lavoisier's oxygen theory of combustion, the replacement of Newtonian mechanics by Einstein's relativity theory, and, throughout the book, the replacement of Aristotle's physics and Ptolemy's astronomy by the Copernican view that the sun is the center of the planets' motions. His account of phlogiston chemistry provided a clear example of the kinds of changes that occurred during scientific revolutions, while his discussion of Einstein permitted a detailed examination of one of his major critical innovations, the concept of incommensurability. But the Copernican revolution proved problematic. It failed to conform to the general pattern of a revolution, preceded by a crisis, which in turn had been generated by an anomaly. Even though Kuhn believed at the time that astronomy in Copernicus' day was a good example of a crisis state (it is not: see Gingerich 1975 and Goldstein 1991), he could not point to an empirical anomaly of the type that he believed had motivated other revolutionary changes. He was therefore left in the ironic situation that his prototype scientific revolution, the Copernican revolution, did not really conform to the pattern that he was sketching for scientific revolutions in general. In this book, we shall argue that the Copernican revolution did precipitate revolutionary changes in the conceptual structure of astronomy, although these changes were not correctly located by Kuhn. We will argue that Copernicus' work can be seen as a minor variation on the conceptual structure in astronomy established by Claudius Ptolemy. Copernicus' work in astronomy, as opposed to cosmology, is not incommensurable with Ptolemy's. The revolutionary break occurs with Kepler, and it introduces not only a new conceptual structure that is incommensurable with the old one, but a new *type* of concept in astronomy.

Kuhn suggested that anomalies created the crises that caused revolutions. But an anomaly is not merely an experimental or observational failure. Rather it is a phenomenon that resists easy interpretation or classification according to accepted knowledge. We shall show that many important anomalies conform to a pattern illustrated as follows. Suppose that all the birds you have ever encountered resemble either chickens or ducks. How do you classify a bird that has the beak of a chicken, but webbed feet? When a bird called a screamer was discovered in South America during the nineteenth century, something very like this actually happened, and as a result the original categories used to classify birds had to be replaced with new and incompatible ones. We shall show how such responses to anomalies can be understood through a cognitive theory of concepts and categorization, and provide the basis for understanding incommensurability and revolutionary change.

1.3 THEORIES OF CONCEPTS

Between 1969 and 1994, Kuhn elaborated an account of scientific change in which the theory of concepts holds a central place. From the very first presentation of his work, Kuhn had introduced ideas that he found in the later writings of Wittgenstein on the nature of concepts and rule following. In developing his own account of concepts he extended Wittgenstein's account of family resemblance concepts. Like Kuhn's work in philosophy of science, Wittgenstein's account of concepts has been almost universally repudiated by professional philosophers in the English-speaking world. Kuhn's appropriation of Wittgenstein's account might have been no more than another footnote to the history of philosophy were it not for simultaneous developments in psychology. At about the same time, a successful revolution in psychology and allied fields – the Roschian revolution – replaced the classical theory of concepts with a range of new accounts that were remarkably similar to the theory Kuhn had developed.

1.3.1. The Classical Theory of Concepts

As we will use it in this book, the classical theory of concepts asserts that the application of a concept can be completely specified by discovering a set of necessary and sufficient conditions that define the objects falling under the concept. These necessary and sufficient conditions will be stated using other concepts and constitute what a philosopher would call the analysis of the concept, or a grammarian its definition. The secondary concepts introduced by the necessary and sufficient conditions specify certain features possessed by all objects falling under the original concept, but absent from objects that do not fall under that concept. In the most extreme case the list of necessary and sufficient conditions may indicate just one feature shared by all objects falling under the concept but absent from objects not falling under it. In more typical cases, the list of necessary and sufficient conditions, however long, may be taken as defining a single complex predicate or property shared by all objects falling under the concept.

Despite its historical durability, the classical theory of concepts is objectionable on practical, philosophical, and empirical grounds.

From a practical viewpoint, the main objection to the theory has been its intractability. It is more than two thousand years since the theory appeared in the works of Plato, but philosophers have failed to produce a single generally agreed analysis of any important concept that completely specifies the necessary and sufficient conditions of its application. Even relatively trivial cases in which such definitions appear possible remain open to challenge. Two favorite examples of concepts that can be completely analyzed by means of necessary and sufficient conditions are 'triangle' and 'bachelor'. However, if 'triangle' is analyzed as 'a plane figure bounded by three sides', what becomes of triangles drawn on the surfaces of spheres or any of the other surfaces investigated in non-Euclidean geometry, beginning in the nineteenth century? If we accept that three-sided figures drawn on spherical or hyperbolic surfaces fall under the concept, can we also accept that figures drawn in a plane but bounded by nonstraight lines are triangles? And can the lines have breaks in them? A supporter of the classical theory might respond by adding new necessary and sufficient conditions to the original ones. A skeptic might respond that there is no visible end to this process.

The concept of a bachelor fares no better. Suppose we attempt to analyze 'bachelor' as 'unmarried adult male'; then as Lakoff (1987) and others have pointed out this definition applies to many instances that we are otherwise reluctant to count as bachelors. Examples include gay men in permanent relationships, and other individuals, such as the head of the Roman Catholic Church, who are not in a position to marry (Coulson 2001). Although it is sometimes claimed as a virtue of the classical theory of concepts that it explains analytic inferences such as "Smith is unmarried; therefore Smith is a bachelor," it may reasonably be objected that this inference is suspect unless we know that Smith is neither gay nor the pope. The same background information that controls our application of the concept in these cases may also operate when we draw inferences, undermining the supposed 'analytic inferences'. What we need is a theory of concepts that incorporates this background information.

Difficulties of the sort just raised for 'triangle' and 'bachelor' may be attributed to the *open texture* of human concepts, an idea introduced by Wittgenstein (1953) and popularized in lectures by Friedrich Waismann (1965). This feature of language follows from the nature of the linkages between instances of concepts in natural languages, called by Wittgenstein *family resemblance*. In a famous example, Wittgenstein argued that many common concepts like 'game' could not be defined by means of necessary and sufficient conditions on the grounds that there was no single, common feature linking all objects falling under the concept. But these examples contribute to a more fundamental point: Wittgenstein argued for the priority of human practices, including linguistic practices, to the rules that may be devised to regulate or define them. The classical theory's necessary and sufficient conditions, introduced in the analysis or definition of a concept, are enforced as rules to determine the application of the concept. But if, as Wittgenstein argues, practices are always prior to rules, no list of rules will completely determine the application of a concept.

Waismann and many others, including Kuhn, were inclined to see the problem as one of the future application of existing concepts. However successful we have been up to the present moment in specifying necessary and sufficient conditions for the application of a concept, there is, on this view, no guarantee that the next instance of the concept we encounter will not violate the norms specified in the analysis adopted so far. The original framers of the definition of a triangle could not foresee the advent of non-Euclidean geometry. But the difficulty is not just the result of new knowledge in mathematics or the sciences. Everyday situations provide evidence against the classical theory of concepts just as much as the novelties encountered in science: a patch of cloth may be a perfectly good triangle to a child learning the concept or an adult making a quilt, even though none of its three sides is a straight line and it will only repose in a plane after it is ironed. What is needed is a theory of concepts that functions equally well inside and outside the sciences.

Although the ideas of family resemblance and open texture became widely known after the publication of Wittgenstein's *Philosophical Investigations* in 1953, the dominant philosophical position in the English-speaking world remained some version of the classical theory. Defenders of classical theories found too many obscurities in Wittgenstein. But also, perhaps because his approach to philosophy strongly discouraged system building, no systematic alternative to the classical theory was articulated on the basis of Wittgenstein's work until Thomas Kuhn began to develop a theory of concepts, based on Wittgenstein's ideas, but informed by detailed studies of historical change in science.

At the same time, but separate from Kuhn's work, radical developments took place within psychology. These developments constitute the third, empirical objection to the classical theory. Beginning in the 1970s psychologists discovered that human concepts display graded structure. Specifically, human subjects readily rate instances of a given concept as better or worse examples of the concept. Before considering the empirical evidence for this important effect, let us briefly consider its implications as a philosophical counterargument to the classical theory. According to the classical theory all instances of a concept are equal. Every instance falls under the concept because it shares the same common features, those specified by the list of necessary and sufficient conditions that analyzes or defines the concept. So, if the classical theory is correct, there is no way to grade instances of a concept as better or worse examples of the concept. However, empirical studies show that human beings actually grade all instances as better or worse examples of the concept. Hence, the classical theory is false, and whatever human beings are doing when they use concepts does not involve lists of necessary and sufficient conditions.

1.3.2. The Roschian Revolution

Beginning in the early 1970s the American psychologist Eleanor Rosch made a series of studies examining the way in which individuals in many different situations and different cultures grouped objects into categories. Like Kuhn, she decided that no account based on category members sharing a single common feature was adequate to the empirical data she was collecting and concluded that analyses of concepts in terms of necessary and sufficient conditions were defective. The most compelling evidence that she gathered initially concerned the graded structure or typicality of concepts. Rosch found that individuals readily classified objects not only as members of particular categories but also as better or worse examples of the category. Rosch and her successors documented judgments of typicality worldwide among human groups as different as stone age tribes from New Guinea and undergraduate students from the United States.

In the judgment of Rosch's subjects, even objects that uncontroversially belonged to a category differed in how well they represented the category. To take a common example, for Westerners the best examples of the concept 'chair' turn out to be the kind we would expect to find at a dining table: they have four legs; a flat, hard seat, a straight back, and probably lack arms. Arm chairs, easy chairs, recliners, bar stools, three-legged stools, and modernist chairs supported on a single, central column are less good examples of the concept. Similar gradations in 'typicality' or 'goodness of example' appear in the case of natural objects. For Westerners a small bird with a sharp beak, a short neck, and a medium-sized body, like a blackbird, starling, or an American robin, is a good example of the concept. Those with longer legs, necks, or beaks are less good examples. For Asians, however, the best examples of 'bird' are likely to resemble ducks, geese, or swans: by contrast with the Western examples they have rounded beaks, long necks, and larger bodies (Barsalou 1992a: 176). Although Rosch demonstrated surprising agreement on typicality phenomena across cultures, for example, in the case of primary colors (writing as E. R. Heider 1972), the example of 'bird' shows that not all cultures agree on the same best example. Other research has shown that typicality may vary between individuals in a given context and in a single individual on different occasions (Barsalou 1987, 1989; Barsalou and Billman 1989).

What is universal, however, is the rating of particular instances of the concept as better or worse examples. We will refer to this phenomenon as the graded structure of a concept.

The existence of graded structure in human concepts has been demonstrated for a wide variety of different conceptual types, but most importantly for natural kinds and artifacts. Rosch originally demonstrated the existence of graded structures in categories for natural kinds like animals, birds, fish, and trees and artifacts like tools, clothing, and furniture. While these studies depended upon manipulating words, she obtained the same results in studies in which her subjects manipulated color samples or simple geometrical shapes. She concluded that both semantic and perceptual categories display graded structure (Heider 1972; Rosch 1973a,b; Rosch and Mervis 1975; Rosch et al. 1976). Other perceptual categories that display graded structure include human facial expressions (Ekman, Friesen, and Ellsworth 1972). At a more abstract level, notable studies established graded structures for categories including phrases used to designate spatial location (Erreich and Valian 1979), and to classify psychiatric conditions (Cantor, et al. 1980). Basic concepts in geometry and arithmetic were shown to display graded structures. Rosch's original work on the simplest geometrical figures was extended to polygons (Williams, Freyer, and Aiken 1977). A study arguing against Rosch's position ironically presented evidence that number concepts have graded structures (Armstrong, Gleitman, and Gleitman 1983; for a discussion see Lakoff 1987: 148–151). Graded structure was also demonstrated in categories that were completely artificial, or natural but completely novel. Homa and Vosburgh (1976) showed graded structure in artificial categories consisting of dot patterns, while Mervis and Pani (1980) showed the same thing for imaginary objects. Finally, Barsalou demonstrated typicality effects in categories that had been freshly constructed ad hoc, or in pursuit of specific short-term goals (Barsalou 1982, 1991).

Graded structure has also been shown to underlie performance across a wide variety of intellectual tasks (Barsalou 1992a: 175–177). Human subjects classify typical examples of a concept more rapidly than less typical or nontypical examples. Graded structure also appears in the operation of human memory: typical instances of a concept are retrieved from memory earlier and more rapidly than less typical instances. Graded structure influences language acquisition; children become linguistically proficient in typical instances of a concept earlier than in atypical ones. There are also typicality effects in deductive and inductive reasoning. When asked to assess the validity of an incomplete deductive argument, human subjects respond more quickly when the missing premise of the argument involves a typical instance. In a classic example of inductive reasoning, subjects are asked to judge whether the next instance of a concept will have a novel property, given that the last instance possessed this property. Again, their estimates of likelihood vary directly with the typicality of the first instance in which the new property is encountered (Rips 1975).

Rosch's results were most readily accommodated by an account of concepts that lacked a single common feature connecting all category members. Early in her research she recognized the similarity between her empirical results and Wittgenstein's philosophical account of the nature of human concepts. Later work in psychology and cognitive science followed Rosch in recognizing that human concepts conform to Wittgenstein's family resemblance account rather than to the classical theory with its necessary and sufficient conditions.

Rosch's results were replicated widely across cultures and within cultures, using natural categories and artificial categories. By the middle 1980s psychologists commonly referred to a Roschian revolution: a revolution in which the classical theory of concepts that prevailed before Wittgenstein had been replaced by a family resemblance account in the light of research stemming from Rosch's empirical findings. However, Rosch's work did not lead to a single consensus view of the nature of human concepts. A variety of different accounts were introduced using different techniques to generate phenomena like graded structures. By the beginning of the 1990s many of these accounts had in turn been superseded when new empirical findings established the existence of structural connections within conceptual systems over and above the graded structures discovered by Rosch.

In his mature work, Kuhn developed a theory of concepts in which individuals acquire basic categories by learning to discriminate similar and dissimilar features of category members (see Chapter 2). This led him to an account of categories and concepts, like that of Rosch, conforming to the pattern that Wittgenstein described as family resemblance, and not the classical view. Concepts learned in this way could not be defined by necessary and sufficient conditions, as there may very well be no single common feature linking all members of a category. And equally, different members of a single linguistic community may employ different features to classify the same objects as members of particular categories successfully. In this book we will use an account of concepts descended from Rosch's pioneering work that incidentally validates many of Kuhn's fundamental insights on the nature of conceptual systems in science and the dynamics of their change. The particular model of concepts we use – frame theory – goes considerably beyond Rosch and Kuhn. In the next section we describe some of the reasons for our choice.

1.3.3. Three Responses to the Roschian Revolution

The discovery of graded structures and allied effects led to the general abandonment of the classical view of concepts in psychology and related disciplines like cognitive science. A variety of new theories of concepts emerged. Among the most important are the families of theories designated prototype theories, exemplar theories, and frame theories.

Beginning with the work of Rosch herself (Rosch and Mervis 1975) many psychologists and cognitive scientists developed prototype models of human concepts. A prototype is an ideal or maximally typical example (which may not exist as a real exemplar) derived by abstraction from the actual instances of the concept. When prototypes for a range of categories have been acquired, new objects are classified by judging their degree of similarity to the prototypes for each category and assigning them to the category with the highest degree of similarity (Barsalou 1992a: 28–29). But the process of abstraction remains obscure: how and why are certain features of instances selected to form the prototype from a potentially infinite class of candidates? This problem is mitigated by Mervis and Rosch's proposal that certain basic level categories are acquired preferentially, and that the features found in their prototypes have obvious roles in orienting organisms in their environment (Mervis and Rosch 1981). The originators of prototype theory may have hoped to develop a general theory of human concepts. Although effective as a way to accommodate graded structure, the prototype theories generally fail to represent knowledge that human categorizers demonstrably possess about category size, variability among instances, and correlations between the features stored in the prototype. As an example of the last of these: everyone knows that a bird with feet like a duck is unlikely to have a beak like a chicken.

A second response investigated the possibility that human beings store not an ideal abstraction of a category, but information about all previously encountered members of a category. They then perform subsequent classifications by comparing new objects with all the known instances in each category, selecting the best fit (Medin and Schaffer 1978; Brooks 1987). Empirical studies show, for example, that people succeed in categorizing a new instance more readily if it strongly resembles a single instance encountered while learning a category rather than weakly resembling a range of instances encountered during the learning process (Barsalou 1992a: 34–36). Exemplar models are effective ways of representing many of the things missing in earlier prototype accounts such as the sensitivity of categorization to size of category, variability among instances, context, and correlations between attributes (Medin 1989). An obvious problem with exemplar models is the potentially vast calculating machinery that must be imputed to the cognitive system when making a categorization. However, Barsalou suggested that an augmented prototype model could be constructed to combine the best features of both earlier views and went on to develop frame theory (Barsalou 1990, 1992b; Barsalou and Hale 1993).

To understand historical change in science we require a theory of concepts that combines the prototype model's ability to represent graded structure with the exemplar model's ability to capture the details of individual cases. It should also provide a means to represent the details of conceptual structures in ways that illuminate detailed historical cases. Barsalou has developed such a theory by generalizing the concept of the frame (presented in detail in Chapter 3). Barsalou's frames represent concepts by means of layers of nodes, where all the nodes represent specific subsidiary concepts. The relations between the layers of nodes are straightforward. A single node representing a superordinate concept (BIRD) is connected to two layers of nodes selected to represent attributes (LENGTH of NECK, COLOR) and values of those attributes (for example, NECK: LONG or SHORT, COLOR: WHITE or BROWN). A particular subordinate concept (SWAN) may then be represented by activating a specific pattern of values (NECK: LONG, COLOR: WHITE). Variations in these activation patterns may be used to generate different subordinate concepts, which may differ in typicality (generating a graded structure). Patterns of selected values represent the prototyes of subordinate concepts. While the prototypes may share a common set of attributes, the subordinate concepts may have no single common feature. There is no restriction on the number of attributes linked to a major concept, or values linked to each attribute. Rather, the structure of a given concept should be determined empirically by the responses of subjects. However, the structures produced by analyzing the responses of experimental subjects will generally be accepted as obvious by native speakers or by proficient members of the group that supports a given conceptual structure.

Two considerations support the application of these techniques to historical cases. First, the empirical research from which contemporary theories of concepts like the frame theory arose has proved very robust across modern human cultures. The effects captured in frame theory represent genuine human universals and should apply as much to sixteenth-century natural philosophers as to twentieth-century university students. Second, humans are capable of effectively emulating the conceptual structures of groups to which they do not belong (Barsalou and Sewell 1984). This suggested to us that it would be legitimate to apply these techniques to historical cases. In historical cases we have supplied what we hope are plausible lists of attributes and values, given the current state of historical knowledge, and always subject to revision in the light of new historical research.

1.4 NATURE AND SCOPE OF THE PRESENT WORK

During the last twenty-five years, philosophers and others have begun turning to cognitive science for new resources to address the intellectual content of science (Nersessian 1984; Giere 1988, 1992, 1994; Gooding 1990; Thagard 1992; Margolis 2002; Nickles 2003). As Nancy Nersessian has pointed out in a series of important studies (1989, 1992a, 1992b, 1995, 1999, 2001, 2003; Nersessian and Andersen 1998), these resources go beyond traditional philosophy of science and allow us, for example, to address questions excluded or mishandled by social constructivist accounts. Similarly, Miriam Solomon's social empiricist epistemology for science, drawing on cognitive science and other fields including history, anthropology, and feminist theory, defines a position that endorses neither traditional philosophy of science nor constructivist sociology of science (Solomon 2001).

The position we seek to define is neither an endorsement nor a simple rejection of the major positions in science studies. Like many later social constructivists, we regard the scientific community as the main actor in scientific change. The conceptual structures we examine are community property; they are preserved and transmitted by groups (see Sections 2.1 and 2.2). However, the same theory that permits the description of group-based conceptual structures also allows the representation of individual conceptual structures and the extent to which they diverge from group norms. This is important in cases in which a single individual, like Johannes Kepler, initiated a major change in an important structure (introducing the concept of an orbit in the conceptual structure of planetary astronomy, Sections 6.4 and 6.5). We are therefore in a position to consider many issues treated in traditional philosophy of science. We regard the history of science *considered historically* as a source of useful knowledge about the nature of science.

In Chapter 2 we present Kuhn's theory of concepts, which is the foundation for his mature account of scientific change. This account of concept acquisition extends Wittgenstein's family resemblance account of concepts. It can readily accommodate phenomena such as graded structure that undermine the classical theory of concepts. On this account, conceptual systems also carry additional types of information, now dubbed 'knowledge of regularities' and 'quasi-ontological knowledge', in addition to their overt content. We begin a discussion of nuclear physics as it developed up to the 1930s as a historical example.

In Chapter 3 we introduce a version of the frame theory of concepts, including some of the empirical work on which it is based. We show how the frame theory can be used to describe family resemblance and graded structure and hence to represent and extend the theory of concepts introduced in Chapter 2. An important subsidiary point is the role of ancillary theories as the basis for constraints between the elements of individual frames.

In Chapter 4 we give a general account of scientific change using
the resources of the frame model. By considering the general principles that govern conceptual structures, we are able to distinguish both
types and degrees of conceptual change in science. We recover the
distinction between normal science and revolutionary science, while
clarifying many subsidiary issues. We also give a detailed account of the
role of anomalies in normal science, again examining concrete historical examples from nuclear physics in the 1930s. By introducing a new
example based on taxonomies used for classifying birds during the
Darwinian revolution, we show that revolutionary changes may occur
without communication failure and indicate some of the resources
on which rational appraisal may be based even for incommensurable
conceptual structures. The chapter concludes with an extended examination of the Noddack case and the discovery of nuclear fission.
We argue that cognitive factors played a central role in the scientific
community's negative response to Noddack, and that in understanding such cases cognitive factors are not eliminable in favor of social
causes.

In Chapter 5 we present a new account of incommensurability based
on the resources of the frame model. We show that incommensurability
is a real historical phenomenon, which admits of degrees that can be
diagnosed and appraised. Using the frame model, we are able to specify
with great precision the conditions under which incommensurability
appears between conceptual systems. In this chapter we introduce an
extended historical discussion of the Copernican revolution, one of
the main examples of incommensurability, but a historical case that
Kuhn never treated successfully.

The examination of the Copernican revolution continues in Chapter 6. Here the tools developed in earlier chapters are applied to
conceptual systems representing the three main options in sixteenth-
century astronomy: the tradition founded by Claudius Ptolemy, the
Averroist critics of Ptolemaic astronomy, and Copernicus. The surprising result is that the three systems are found to be commensu-
rable. We argue that the conceptual system of Johannes Kepler introduced the first major incommensurability with the ancient astronomical tradition during the Copernican revolution. The analysis enables
us to locate, in a preliminary way, some of the major intellectual
issues that motivated scientific change in astronomy and cosmology

throughout this period. A central feature that has received little attention is the abolition of the celestial orbs. An important general consequence of our analysis is the recognition that the changes producing major incommensurabilities may occur in small increments over an extended period, rather than by all-or-nothing change that takes place instantaneously.

The majority of concepts examined throughout this book are object concepts, for example, those representing what philosophers and grammarians call 'natural kinds'. But many important scientific concepts are event concepts, which differ in structure from object concepts. The replacement of an established object concept by an event concept may mark an especially significant moment in the history of science, a point illustrated by Kepler's introduction of the concept 'orbit'.

In the final chapter we consider the location of our work among the various contemporary options in science studies. We begin by examining one of the main issues separating philosophical accounts of science from social constructivist accounts: realism. We attempt an initial articulation of a new realist position that neither simply endorses nor condemns previous positions, but tries to do justice to the ideas of revolutionary change and incommensurability as they are supported by cognitive studies. The central idea of this viewpoint is the role that the historical state of a field, and especially its conceptual structure, plays in the training of new entrants to the field and in the corresponding definition of new research problems. Having made it clear that we reject the extreme antirealism sometimes expressed in constructivist sociology of science, we next consider how our work stands up to the constructivist critique of traditional philosophy of science. We accept the criticisms implicit in Bloor's celebrated statement of four adequacy conditions for any account of science: causality, symmetry, impartiality, and reflexivity, if read as an attack on the historical standards accepted by philosophers of science. However, we argue that the account presented in this book satisfies all four conditions and is superior to social constructivist accounts of science in its ability to explain historical examples like the Noddack case and the Copernican revolution, in which cognitive factors of different kinds were key elements in determining the historical outcome.

We conclude that cognitive factors must play a role in any general account of historical change in science. Frame theory provides access to these factors in a way that has not been available before. Our results show the permanent value of the ideas of revolution and incommensurability, which deserve a central place in the thinking of every philosopher, historian, and sociologist of science.

2

Kuhn's Theory of Concepts

After the appearance of *The Structure of Scientific Revolutions* in 1962 Kuhn attempted to develop a Wittgensteinian account of family resemblance concepts for a domain that other philosophers had found most unlikely: scientific concepts. In the 1970 postscript to *Structure* Kuhn even suggested that the variants of Newton's second law that applied to different physical systems showed family resemblance, but no single defining properties. If this proved to be the general case, the most important examples of scientific concepts might turn out to be family resemblance concepts rather than the well-behaved concepts, analyzable by necessary and sufficient conditions, expected by earlier philosophers of science. Kuhn returned to these themes again and again in his later philosophical writings.

Kuhn's theory of concepts focused on a restricted class of terms, namely, kind terms. As Kuhn defined *kind terms* they are "primarily the count nouns together with the mass nouns, words which combine with count nouns in phrases that take the indefinite article. Some terms require still further tests hinging, for example, on permissible suffices" (Kuhn 1991: 92). Thus, kind terms include natural kinds, artifactual kinds, and social kinds.

2.1 EXEMPLARS

An important source for Kuhn's theory of concepts was his early reflection upon science teaching. He made the observations that, first,

19

science education is based entirely on prepared teaching materials, and, second, this teaching confers the ability to recognize resemblances between novel problems and problems that have been solved before. This ability seems to rely almost exclusively on exemplary problems and concrete solutions rather than on abstract descriptions and definitions. Textbooks do not describe the sort of problems that the discipline deals with in the abstract, but "exhibit concrete problem solutions that the profession has come to accept as paradigms, and they ask the student, either with a pencil and paper or in the laboratory, to solve for himself problems very closely related in both method and substance to those through which the textbook or the accompanying lecture has led him" (Kuhn 1959/1977: 229). Within linguistics the term 'paradigm' is used to denote conjugation patterns, such as the pattern displayed by the Latin verb *amo, amas, amat, amamus, amatis, amant.* Kuhn claimed that the procedure by which science students are supposed to model novel problems on exemplary problems is similar to the procedure by which language students learn conjugations by extracting patterns from examples. He adopted the term 'paradigm' to denote standard examples in science teaching; thus, that term first entered Kuhn's work prior to the publication of *The Structure of Scientific Revolutions* to denote standard scientific problems, or *exemplars,* used in teaching.

2.2 THE LEARNING PROCEDURE

Kuhn's theory of concepts starts from the claim that exemplars are the vehicle of science teaching. Illustrating his theory with an example of how a child learns to recognize waterfowl, Kuhn gradually developed a full-fledged family resemblance theory of concepts.

According to this theory, the basic conceptual structure of science is a classification system that divides objects into groups according to similarity relations. The grouping is not determined by identifying necessary and sufficient conditions, but by learning to identify similarities and dissimilarities between the objects. It was one of Kuhn's central claims that one learns such concepts by being guided through a series of encounters with objects that highlight the relations of similarity and dissimilarity currently accepted by a particular community of concept users. In this process, learning depends upon examining similar or

FIGURE 1. Instances of waterfowl. Reproduced from Thomas S. Kuhn, "Second thoughts on paradigms," in F. Suppe (ed.), *The Structure of Scientific Theories* (Urbana: University of Illinois Press, 1974), p. 476. Copyright 1977 by Board of Trustees of the University of Illinois. Used with permission of the University of Illinois Press.

dissimilar features of a range of objects (Kuhn 1974, 1979; see also Hoyningen-Huene 1993: Ch. 3.6).

Kuhn's standard example of a learning process of this sort is a child learning the concepts 'duck', 'goose', and 'swan' (Kuhn 1974). In this example, an adult familiar with the classification of waterfowl guides a child ("Johnny") through a series of ostensive acts until he learns to distinguish ducks, geese, and swans. Johnny is shown various instances of all three concepts being told for each instance whether it is a duck, a goose, or a swan (Figure 1). He is also encouraged to try to point out instances of the concepts. At the beginning of this process he will make mistakes, for example, mistaking a goose for a swan. In such cases Johnny will be told the correct concept to apply to the instance pointed out. In other cases he ascribes the instance pointed out to the correct concept and receives praise. After a number of these encounters Johnny has, in principle, acquired the ability to identify ducks, geese, and swans as competently as the person instructing him (Figure 2). Although the similarity classes for 'duck', 'goose', and 'swan' may be imagined to have homogeneous members, a moment's thought

FIGURE 2. The similarity classes of ducks, swans, and geese. Reproduced from Thomas S. Kuhn, "Second thoughts on paradigms," in F. Suppe (ed.), *The Structure of Scientific Theories* (Urbana: University of Illinois Press, 1974), p. 476. Copyright 1977 by Board of Trustees of the University of Illinois. Used with permission of the University of Illinois Press.

will show that the members of these classes – individual ducks, for example – bear no more than a family resemblance to each other.

During the ostensive teaching, Johnny encounters a series of instances of the various waterfowl and examines these instances in order to find features with respect to which they are similar or dissimilar. In this learning process, "the primary pedagogic tool is ostension. Phrases like 'all swans are white' may play a role, but they need not" (Kuhn 1974: 309). In this way a conceptual structure is established by grouping objects into similarity classes corresponding to the extension of concepts. It is an important feature of Kuhn's account that this grouping can be achieved solely by learning to identify similarities between objects within a particular similarity class and dissimilarities to objects ascribed to other similarity classes. Hence, simple categories like 'duck', 'goose', and 'swan' may be transmitted from one generation to the next solely by extracting similarity and dissimilarity relations from the exemplars on exhibit.

During the subsequent development of his position Kuhn elaborated the emphasis on exemplars rather than abstract rules. His argument had developed from observations on science education and may seem to concern only the novice. Turning to the expert one might think that although scientists had once learned to identify scientific problems by resemblance to exemplars rather than by rules, they might well have abstracted rules for themselves later. However, Kuhn found little reason to believe this. His skepticism was grounded in both contemporary and historical observations. First, scientists "are little better than laymen at characterizing the established bases of their field, its legitimate problems and methods" (Kuhn 1970a: 47). Not only are very few abstract rules to be found in science texts, but even "if asked by a philosopher to provide such rules, scientists regularly deny their relevance and thereafter sometimes grow uncommonly inarticulate" (Kuhn 1974: 305). Second, if one studies the history of scientific research one will note "the severe difficulty of discovering the rules that have guided particular normal-science traditions" (Kuhn 1970a: 46).

Kuhn was making a general claim about the whole of scientific practice when he maintained that the link between research problems within a given discipline "is not that they satisfy some explicit or even some fully discoverable set of rules and assumptions that gives the tradition its character and hold upon the scientific mind. Instead, they may relate by resemblance and by modelling to one or another part of the scientific corpus which the community in question already recognized as among its established achievements" (Kuhn 1970a: 45–46).

Kuhn adopted Wittgenstein's notion of family resemblance concepts, suggesting that research based on exemplars may share a family resemblance rather than being related by specific rules of methodology. This became still clearer in the 1970 postscript, in which Kuhn suggested that the variants of Newton's second law applicable to different physical systems shared a family resemblance, but no single defining common properties (Kuhn 1970a: 188ff.). Here Kuhn pointed out that the basic form of Newton's second law

$$\mathbf{F} = m\mathbf{a}$$

transforms into different but similar forms for different kinds of problem situations:

free fall $mg = m(d^2s/dt^2)$

simple pendulum $mg\sin\theta = -ml(d^2\theta/dt^2)$

interacting harmonic oscillators $m_1(d^2s_1/dt^2) + k_1s_1 = k_2(s_2 - s_1 + d)$

This example, using a fundamental principle of classical mechanics, suggested that the most important instances of scientific concepts might turn out to be family resemblance concepts rather than concepts definable by necessary and sufficient conditions, as expected in the philosophical tradition. Subsequently, Giere (1994) has vindicated Kuhn's controversial claim with a detailed analysis of the family resemblance structure of classical mechanical models.

2.3 SIMILARITY, DISSIMILARITY, AND KIND HIERARCHIES

Dissimilarity plays as important a role as similarity in establishing similarity classes. Attempting to determine a category by similarity alone would be attempting to determine the category solely by what the objects have in common, and there are several reasons why such an attempt would fail.

First, such an attempt would fail for the very same reasons that necessary and sufficient conditions fail. Different pairs of members of the same category may very well have different things in common. Second, a standard objection against categorizations based exclusively on similarity points out that since we can always find *some* similarity between instances of one concept and those of another, similarity alone does not suffice to limit the extension of concepts. In sum, since different pairs of members of the same category may have different things in common, and some members may have some of these things in common with members of other categories, similarity alone clearly does not suffice to determine a category.

Kuhn recognized these problems (Kuhn 1974: 307; similarly Kuhn 1970a: 200) and suggested that they could be solved by including among a concept's constitutive relations not only similarities between members of the same class, but also dissimilarities to members of other classes:

> Note that what I have here been calling a similarity relation depends not only on likeness to other members of the same class but also on differences from the members of other classes.... Failure to notice that the similarity relation appropriate to determination of membership in natural families must be triadic rather than diadic has, I believe, created some unnecessary philosophical problems. (Kuhn 1976: 199)

Hence, on Kuhn's account, the relations of dissimilarity serve to separate members of the category in question from objects in other

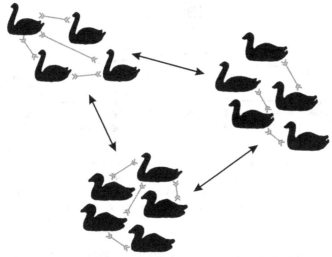

FIGURE 3. Similarity classes are established by both similarity between members of the same class and dissimilarity between members of contrasting classes.

categories to which they could otherwise mistakenly have been assigned (Figure 3). The dissimilarity relation that Kuhn introduced here is not a relation between the instances of arbitrary pairs of concepts, but a relation between instances of concepts in a *contrast set*, that is, a set of nonoverlapping concepts that are all subordinates to the same immediate superordinate (see Kuhn 1983a: 682; 1991: 4; 1993: 317ff.). For example, the concepts 'duck', 'goose', and 'swan' are all subordinates to the superordinate concept 'waterfowl'. This concept is also a family resemblance concept whose instances resemble each other more than they resemble members of contrasting categories such as 'songbird' and 'game bird'. Kuhn's emphasis on the importance of dissimilarity relations therefore serves to avoid the problem that instances of different but highly similar categories might be mistaken for each other and leads to the view that contrasting concepts must always be learned together: "Establishing the referent of a natural-kind term requires exposure not only to varied members of that kind but also to members of others – to individuals, that is, to which the term might otherwise have been mistakenly applied" (Kuhn 1979: 200). Kuhn's restriction of dissimilarity to instances of concepts forming contrast sets can also be found in other fields, such as cognitive psychology (Rosch 1987:157) or ethnographic semantics and

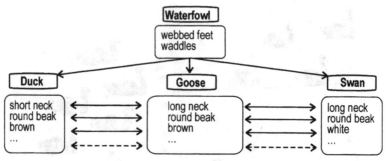

FIGURE 4. Contrast set constituted by similarity and dissimilarity relations represented as a kind hierarchy. Double-headed arrows indicate mutually exclusive properties: if an object instantiates any one property, then it cannot instantiate another property linked by double-headed arrows.

cognitive anthropology (Conklin 1969; Kay 1971). We shall return in more detail to the characteristics of contrast sets in Section 4.2.

Obviously, this analysis can be extended to new superordinate and subordinate levels. Just as the superordinate concept 'waterfowl' can be divided into the contrasting subordinates 'duck', 'goose', and 'swan', so too each of the subordinate concepts can be further subdivided into the particular species of ducks or geese or swans. The hierarchical conceptual structure that arises is one in which a general category decomposes into more specific categories that may again decompose into yet more specific categories, in other words, a *kind hierarchy* (Figure 4). Drawing on the dissimilarity between members of contrasting concepts, family resemblance therefore becomes tied to kind hierarchies. Kuhn never stated this argument explicitly, but only noted, "[A] fuller discussion of resemblance between members of a natural family would have to allow for hierarchies of natural families with resemblance relations between families at the higher level" (Kuhn 1970b: 17, fn. 1). Some features hold for all concepts in the contrast set; others may differ for different concepts in the contrast set. Each similarity class is constituted by a network of overlapping and crisscrossing relations of similarity and dissimilarity. We shall return to a more advanced representation of contrast sets in Chapter 3.

Kuhn also realized that the problem that anything is similar to anything else in some respect would only be solved by the use of contrast sets if the dissimilarity relations between objects were of a specific kind. Kuhn admitted that if the chains of similarity relations developed gradually and continuously, it would indeed be necessary to define where the extension of one concept ended and the extension of the contrasting concept began: "Only if the families we named overlapped and merged gradually into one another – only, that is, if there were no *natural* families – would our success in identifying and naming provide evidence for a set of common characteristics corresponding to each of the class names we employ" (Kuhn 1970a: 45). He therefore argued that the possibility of classifying objects into family resemblance classes depends on an "empty perceptual space between the families to be discriminated" (Kuhn 1970a: 197, fn. 14; similarly Kuhn 1974: 508ff.). On this point, Kuhn explicitly claimed to have moved beyond Wittgenstein: "Wittgenstein . . . says almost nothing about the sort of world necessary to support the naming procedure he outlines" (Kuhn 1970a: 197, fn. 14).

The condition that there must be an empty perceptual space between the natural families to be discriminated may be taken to imply entity realism. On such an interpretation the world comes divided into natural families, and similarities and dissimilarities might therefore simply be read off the world itself. However, this is an interpretation that Kuhn denied. Rather, there may be empty perceptual spaces along many different dimensions, and although some of them may be compatible, they need not all be so. Instead, empty perceptual spaces may separate different, overlapping families along different dimensions in perceptual space. Hence, the condition that there must be empty perceptual spaces between the categories does not imply that there is only one, true categorization based on similarity and dissimilarity in specific respects that can be read off the world itself. Instead, there may be different categorizations based on similarity and dissimilarity with regard to different (sets of) features. In the history of science such different categorizations may give rise to incommensurable positions, for example, Ptolemaic astronomy and Keplerian astronomy, or nuclear physics before and after the discovery of nuclear fission. We shall return to the issue of realism in Chapter 7.

2.4 KNOWLEDGE OF ONTOLOGY AND KNOWLEDGE OF REGULARITIES

The condition that there must be empty perceptual space between the categories means that it can be assumed that in a kind hierarchy

all objects fall clearly into one of the categories. Objects not belonging to any of the known similarity classes are simply assumed not to
exist. Hence, possession of a conceptual structure implies assumptions of what exists and what does not exist, in other words, ontological knowledge. However, different conceptual structures based on
different similarity and dissimilarity relations may well imply different assumptions about what exists and what does not exist. This kind
of ontological knowledge arises from particular individual conceptual
structures, which may have alternatives, and it may therefore be termed
quasi-ontological knowledge.

Through the relations of similarity and dissimilarity constituting a
kind hierarchy, possession of a language implies expectations of the
different situations that nature does and does not present (see Kuhn
1970a; 191). This knowledge of how nature behaves is made possible
because the concepts are not explicitly defined, and different members
of a single linguistic community may focus on quite different features
of the objects they categorize, but nevertheless achieve identical group-
ings. Their ability to categorize objects successfully on the basis of different characteristics is evidence for an empirical correlation among
these characteristics. These empirical correlations between characteristics may be used to predict additional characteristics of an object that
we recognize because it possesses some minimal set. For example, a
bird that has webbed feet probably does not build its nest in a tree.
Hence, apart from ontological knowledge the conceptual structure
also implies *knowledge of regularities.*

The notions of quasi-ontological knowledge and knowledge of regularities in this context have been introduced by Hoyningen-Huene
(1993: Ch. 3.7). It is important to note that both are implied by the
relations of similarity and dissimilarity. Further, since the objects in a
similarity class bear no more than a family resemblance to each other,
there are no restrictions on *which* characteristics can be used when
judging objects similar or dissimilar. Nothing limits the possible characteristics according to which the objects may be similar or dissimilar.
On the contrary, anything one knows about the referents can be used
when matching them with terms. As Kuhn states it: "In matching terms
with their referents, one may legitimately make use of anything one
knows or believes about those referents" (Kuhn 1983a: 681). However,
this means that there is no distinction between the characteristics that

language users may employ to identify an instance of a given concept and the characteristics of an identified instance of a concept that pertain to instances of this concept only on empirical grounds. Consequently, there is no basis for making a distinction between what is entailed in knowing that a given class of objects exists and in knowing how the members of this class behave. Therefore, in introducing the notions of knowledge of regularities and ontological knowledge, Hoyningen-Huene emphasized that "both epistemic components are mutually inextricable moments of the knowledge contained in immediate similarity relations" (Hoyningen-Huene 1993: 112, similarly 117).

This highly abstract discussion may be clarified by returning to the example of waterfowl. In Figure 4, ontological knowledge of waterfowl is expressed through the scope of the contrast set: objects not belonging to any of the known similarity classes are assumed not to exist. Hence, ducks, geese, and swans are expected to exist, but not duckgeese or gooseswans. Knowledge of regularities regarding waterfowl is implied by the correlation of characteristics for each concept in the contrast set. Hence, once a duck is identified by its brown color and round beak, the language user may know without prior investigation that the duck will have webbed feet and waddling gait. However, knowledge of regularities and ontological knowledge are two inseparable aspects of the knowledge contained in concepts; there is no distinction between the features that can be used to identify an object and the features expressing further empirical knowledge about this object.

2.5 INDIVIDUAL DIFFERENCES AND GRADED STRUCTURES

For Kuhn, in contrast to the traditional view, there is no distinction between defining and contingent features of an object. Although different speakers use a given concept to pick out the same similarity class, they need not identify instances and noninstances by the same features. The child learning to distinguish ducks, geese, and swans may identify them by a combination of their colors and beak shapes. The teacher may use beak shape and length of neck. A third person might use length of neck and body size, or any other feature that distinguishes ducks from geese and swans. In principle, different individuals may use totally different features to identify instances of the same similarity class.

Also contrary to the classical view, the family resemblance view implies that different instances of the same concept are not necessarily equally good examples of this concept. As explained in Chapter 1, the classical view implies that all objects falling under a concept do so in virtue of sharing the same list of features, and, hence, all are equal as instances of the concept. On the family resemblance view, on the contrary, category membership is determined from the degree of similarity and difference based not on a fixed set of features, but on a set of features that may vary. Hence, instances of a concept that are similar with regard to many features may be considered better examples of the concept than instances that share only a few of these features. Similarly, instances of a concept that differ from instances of contrasting concepts with respect to many features may likewise be considered better examples of the concept than instances that differ by only a few features. Thus, it is a direct consequence of the family resemblance view that concepts have graded structures.

When Kuhn emphasizes that different speakers may judge category membership from different features, it follows that they may also differ in their judgments about how typical the same instance is or how good an example of the concept this instance is. Hence, it is also a consequence of this model that different speakers may develop different graded structures for the same concept, as described previously in Section 1.3.2.

2.6 GENERALIZATION TO SCIENTIFIC CONCEPTS

For most of his insights Kuhn relied on the example of the child learning the concepts 'duck', 'goose', and 'swan' to illustrate his theory of concepts. Although his aim was to develop a theory of scientific concepts, he deliberately chose to rely on an example from everyday language because the former would "prove excessively complex" (Kuhn 1974: 309). The only example of the acquisition of scientific concepts that Kuhn spelled out in some detail is his analysis of how students learn the concepts 'force', 'mass', and 'weight' (Kuhn 1989: 15–21; 1990: 301–308). However, Kuhn maintained that, in principle, advanced scientific concepts are acquired by the same similarity-based process as everyday concepts: "The same technique, if in a less pure form, is essential to the more abstract sciences as well" (Kuhn 1974: 313). Johnny was presented with various waterfowl and told whether they were ducks, geese, or swans, but science students are presented with a problem situation after first being shown the appropriate expression of a law sketch, $F = ma$ in Kuhn's example, that can be used to solve the problem. Next, the students are presented with further problem situations and must try to assign the appropriate expression for themselves. In this process, the students examine the problems in order to find features with respect to which they are similar or dissimilar. Thus, in learning scientific concepts the student is presented with a variety of problems that can be described by various forms of a law sketch. In this process, the student discovers a way to see each problem as *like* a previously encountered problem. Recognizing the resemblance, the student "can interrelate symbols and attach them to nature in the ways that have proved effective before. The law sketch, say $F = ma$, has functioned as a tool, informing the student what similarities to look for, signalling the gestalt in which the situation is to be seen" (Kuhn 1970a: 189). A conceptual structure is established by grouping problem situations into similarity classes corresponding to the various expressions of the law sketch.

2.7 NOMIC AND NORMIC CONCEPTS

At this point it is important to note a limitation of Kuhn's account acknowledged by Kuhn himself. In his later work Kuhn suggested that some important scientific concepts were not acquired by the processes discussed so far, that is, through learning similarity and difference relations by ostension, leading to the formation of contrast sets that can be represented by concepts forming a kind hierarchy. To mark this distinction, he introduced the term 'normic' to designate concepts acquired by means of these processes. A second class of concepts, prominent in scientific laws, Kuhn labeled 'nomic' (Kuhn 1993). As the reader will understand in subsequent chapters, our main interests lie in extending the account of concepts offered by Kuhn using techniques from cognitive psychology. Like Kuhn, we would regard it as hubristic to claim a complete account of all scientific concepts. However, something needs to be said about the possibility that nomic concepts represent a second and distinct category of scientifically important concepts.

The distinction that Kuhn introduced between normic and nomic concepts is based on whether or not the generalizations in which the concepts appear are exceptionless. For normic concepts such as 'liquid', 'gas', or 'solid' there may be exceptions to the generalization usually satisfied by their referents. For example, the generalization "Liquids expand when heated" fails for water between 0 and 4 degrees Celsius (Kuhn 1993: 316). For nomic concepts such as 'force' or 'mass', on the contrary, the generalizations satisfied by their referents are exceptionless laws of nature. For example, the nomic concept 'force' is involved in the generalizations expressed through Newton's three laws of motion, and they are all exceptionless.

This distinction between normic and nomic concepts resembles a distinction that Kuhn had previously introduced, between concepts applied by direct inspection and concepts for which laws and theories enter into the process of establishing reference (Kuhn 1979: 200). Concepts of the former kind, concepts applied by direct inspection, are concepts like 'duck', 'goose', and 'swan', which are acquired together in contrast sets on the basis of similarity and difference between instances. These concepts, which Kuhn also called basic terms, are thus learned through ostension of individual instances of the concepts in question.

Concepts of the latter kind, concepts for which laws and theories enter into the establishment of reference, are learned by having problem situations pointed out to which a given law applies. For example, to acquire the concept 'force' one may have pointed out problem situations to which Newton's second law applies such as the simple pendulum, free fall, or the harmonic oscillator. On this view, problem situations form similarity classes in much the same way as instances of basic concepts do. The difference between normic and nomic concepts is therefore not a distinction between concepts based on similarities and concepts that can be explicitly defined, but rather a difference between the level at which the similarities enter.

However, there is a further difference between normic and nomic concepts. Whereas for normic concepts like 'duck' and 'goose' several instances of each individual concept in the contrast set will be ostended to the language learner, for nomic concepts what is pointed out are not instances of individual concepts but complex problem situations to which a given law applies and that involve the simultaneous use of several nomic concepts. Instances of normic concepts like 'duck', 'goose', and 'swan' are ostended individually, and in this process each individual object is ascribed to one of the concepts in a contrast set and simultaneously not ascribed to the other concepts in the set. For nomic concepts, in contrast, it is the instances of the application of a natural law that are pointed out. For example, in Newton's second law $\mathbf{F} = m\mathbf{a}$, the concepts 'force', 'mass', and 'acceleration' appear together in all ostended problem situations.

Kuhn never offered a general account of how to identify the reference of the individual nomic concepts that appear together in problem situations to which a given natural law applies. However, the similarity view does apply to problem situations that cannot be defined but exhibit a set of family resemblances. Consequently, the representation of nomic concepts requires two layers: one that represents the similarities and differences between complex problem situations involving more than one concept and another layer representing the salient features of these concepts individually (Andersen and Nersessian 2000). Having noted this possible limitation of the account we are developing, we return to Kuhn's main concern, the consideration of normic concepts, in the next section of this chapter.

2.8 A SCIENTIFIC CONCEPTUAL STRUCTURE: EARLY NUCLEAR PHYSICS

As an example of a kind hierarchy with similarity classes based on several characteristics we shall examine the kind hierarchy of radioactivity. First, we shall briefly examine the initial development of this field of inquiry, from the discovery of 'uranium rays' until this phenomenon was differentiated into three kinds of radioactivity, α decay, β decay, and γ decay. This is not intended as a complete historical account (for detailed accounts see Kragh 1999; Pais 1986). Next, we shall examine in more detail how a few decades later a particular line of research investigated the kind of decay that resulted from bombarding uranium, by then the heaviest known element, with neutrons.

The development of this field of inquiry began with the discovery of x-rays. In 1895 Röntgen had discovered that some kind of rays

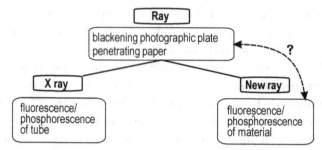

FIGURE 5. Becquerel's proposal for a new correlation of features.

were emitted when cathode rays reached the end of a Geissler tube and made the glass fluoresce. These new rays would make a specially coated screen fluoresce or blacken a photographic plate. The rays penetrated paper, wood, a thick aluminum layer, and various other materials. Röntgen termed the new phenomenon x-rays (Kuhn 1970a: 57–59; Röntgen 1896). Soon afterward, the discovery was discussed in the French Academy of Sciences. Poincaré suggested that the rays might be related to the fluorescence at the end of the Geissler tube and that other fluorescing bodies might also emit x-rays (Poincaré 1896). At three subsequent meetings in February 1896, Henry, Niewenglowski, and Becquerel reported to the French Academy of Sciences about the experiments they had conducted on whether fluorescing bodies in general emitted x-rays (Henry 1896; Niewenglowski 1896; Becquerel 1896a). For example, Becquerel had conducted an experiment in which he placed a fluorescent uranium salt on a photographic plate that was wrapped in black paper. Having exposed the uranium salt to sunlight so that it became fluorescent, he observed a blackening of the photographic plate (Becquerel 1896a).

There was, as yet, no model that explained *why* the features of fluorescence and radiation should be correlated; there was only the observation of the correlation itself (Figure 5). But by accident Becquerel was able to examine the circumstances in which this correlation between features would hold when he repeated the experiment; because of the lack of sunlight the uranium salt was not fluorescent, yet the photographic plate was still blackened (Becquerel 1896b). Apparently, the new phenomenon was not correlated with fluorescence. Becquerel soon discovered another feature correlation when he found that the rays would discharge an electroscope (Becquerel 1896c), and later that various minerals containing uranium emitted some sort of

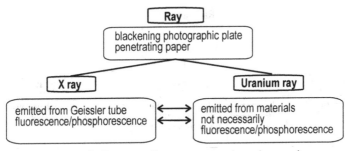

FIGURE 6. Becquerel's new concept 'uranium ray'.

radiation, even if the minerals were *not* fluorescent. Hence, starting by analogy to an already existing concept – x-rays – Becquerel (1896d) formed a new concept termed *uranium rays* (Figure 6). But uranium rays were not triggered by or dependent upon fluorescence; they were simply intrinsic to the material that emitted them. Because uranium rays differed in origin from x-rays, two main branches were formed in the kind hierarchy.

At first, Becquerel believed that the uranium rays were only emitted by uranium compounds, but in 1898 Marie Curie discovered that thorium also emitted the new kind of rays (Curie 1898). The new rays were therefore not uniquely associated with uranium, and the Curies started calling them Becquerel rays rather than uranium rays, while referring to the minerals emitting the rays as radioactive substances. Examining the increase of radioactivity with the increase of uranium present, the Curies also discovered that pitchblende and calcite were much more radioactive than their uranium content would indicate. They hypothesized that the two minerals contained another element that would be more radioactive than uranium and named it polonium (Curie and Curie 1898). A few months later they found that pitchblende contained yet another highly radioactive element, which they named radium (Curie, Curie, and Bémont 1898). What had started as 'uranium rays' in analogy to x-rays would soon form a kind hierarchy of various kinds of radiation.

In 1899, on the basis of a series of absorption experiments, Rutherford showed that uranium rays were complex and contained at least two distinct types of radiation, "one that is very readily absorbed, which will be termed for convenience the α radiation, and the other of a more penetrative character, which will be termed the β radiation" (Rutherford 1899: 175). In 1900, Villard discovered γ radiation when

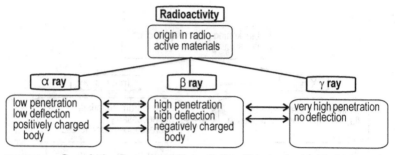

FIGURE 7. Correlation between differentiating features in the kind hierarchy of 'radioactivity'.

he found a very penetrating kind of radiation that was not deflected in a magnetic field (Pais 1986: 62). The deflection of the different kinds of rays in a magnetic field had been studied in different laboratories, and in 1903 Rutherford gave the following summary:

Radium gives out three distinct types of radiation:

(1) The α rays, which are very easily absorbed by thin layes of matter, and which give rise to the greater portion of the ionization of the gas observed under the usual experimental conditions.

(2) The β rays, which consist of negatively charged particles projected with high velocity, and which are similar in all respects to cathode rays produced in a vacuum tube.

(3) The γ rays, which are non-deviable by a magnetic field, and which are of a very penetrating character. (Rutherford 1903: 549)

He added that α rays are deviable by a strong magnetic and electric field, and that they must consist of positively charged bodies. In this way additional differentiating features were correlated with those already found, and this correlation further stabilized the kind hierarchy (Figure 7). At this point, it was still open whether the kind hierarchy was now exhaustive or whether still further differentiations were to come. For example, Blondlot postulated the existence of N rays in 1903, but after careful experiments their existence was rejected by the majority of physicists (Kragh 1999: 35; Nye 1980).

In his description of this initial development of research on radioactivity, Kragh emphasizes that "the early work in radioactivity was primarily experimental and explorative. Which substances were radioactive?

How did they fit into the periodic system of the chemical elements? What were the rays given off by the radioactive bodies? Was the activity affected by physical or chemical changes? These were some of the questions that physicists addressed around the turn of the century" (Kragh 1999: 32). In our reconstruction it becomes clear how this initial explorative research is focused on empirical examination of various possible correlations of features: which features seem to be correlated, which new concepts based on these feature correlations arise, and so on. The development of *reasons* for these correlations, theories that would explain *why* specific features were correlated, did not occur until later. This fits well with Pais' observation that "in those days, theoretical physicists did not play any role of consequence in the development of this subject, both because they were not particularly needed for its descriptive aspects and because the deeper questions were too difficult for their time" (Pais 1977: 927). This also explains why it remained open whether the kind hierarchy of α, β, and γ rays was exhaustive, or whether still other kinds of rays might exist. Without detailed theoretical explanations of which features were correlated and which were not, few correlations could be ruled out. Many different possible combinations of features might exist to form different categories, and only careful empirical investigations would show whether instances could actually be found or not.

By 1902 it had become clear that β rays are electrons (Pais 1986: 87). In 1905 Ramsay and Soddy discovered that helium was produced in the transformation of radium, and in 1908 Rutherford and Geiger concluded from their experiments that α rays consist of helium atoms that had gained a positive charge (Pais 1986: 60ff.). After some initial speculations that γ rays were a particularly powerful form of β rays, in 1913 it was finally established that they were instead the same kind of phenomenon as x-rays (Pais 1986: 62).

Models of α and β radiation explained the correlations of features for each concept in the contrast set – they were particles with specific weights and specific charges, and this explained both their deflection in a magnetic field and their penetrating power in various materials. In Chapter 4 we shall return to the role of models in the explanation of feature correlations. But first, we shall investigate a later episode in the development of nuclear physics that illustrates the role of models in determining the range of a contrast set.

By 1929, Gamow suggested that the nucleus should be seen as a droplet that consisted of individual α particles, protons, and electrons (Gamow 1929a, 1929b). Like a droplet, the nucleus was held together by surface tension. Thus, it was the surface tension that determined which particles or conglomerates of particles could escape the nucleus. Gamow's liquid drop model was used to create a mathematical model of nuclear decay. The core of the mathematical model was the potential well created by the short-ranged attractive forces between the particles inside the nucleus. From the energy function, the probability of a particle tunneling through the potential barrier could be computed, and these computations showed that only particles up to the size of the α particle could be emitted from the nucleus. In this way, the model could explain the scope of the kind hierarchy: the kind hierarchy was exhaustive because particles bigger than the α particle could not tunnel through the potential barrier.

Gamow's liquid drop model became very popular, and the result that only particles up to the size of the α particle could be emitted led to diagrams like checkerboards depicting possible disintegrations. These diagrams were purely descriptive; they did not serve any explanatory purposes, but they represented the scope of the kind hierarchy very well. By using checkerboard-like diagrams to represent nuclear disintegrations, the only processes that could be represented were one nucleus transforming into another nucleus nearby in the periodic table by the emission of a small particle (Figure 8). These diagrams could not represent one nucleus splitting into two much smaller nuclei. Hence, as a simple tool, the diagrams helped consolidate the kind hierarchy.

Gamow had started investigating spontaneous disintegrations of nuclei, but in his 1931 book *Constitution of Atomic Nuclei and Radioactivity* – the first textbook on nuclear physics as such – he also covered artificial transformations of nuclei by collision with α particles. In 1934, Irene Curie and her husband, Frédéric Joliot, discovered that such bombardments could in some cases lead to the production of unstable nuclei that would create stable nuclei through subsequent decay (Curie and Joliot 1934). Two years previously, the neutron had been discovered. Fermi's group in Rome started investigating which elements could be activated by neutron bombardment and how they decayed. A year after Curie and Joliot's publication, in their book

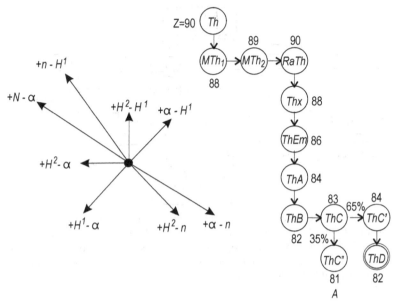

FIGURE 8. Checkerboard diagrams of possible nuclear disintegrations. From Gamow 1931; Meitner and Delbrück 1935.

Der Aufbau der Atomkern: Natürliche und künstliche Kernumwandlungen (1935) Meitner and Delbrück summarized the situation. Only three kinds of processes had been found:

$$_Z^M A + {}^1 n \to {}_{Z-2}^{M-3} A + \alpha$$
$$_Z^M A + {}^1 n \to {}_{Z-1}^{M} A + p$$
$$_Z^M A + {}^1 n \to {}_{Z}^{M+1} A$$

where M and Z are whole numbers denoting the atomic mass, and the atomic number (or electric charge of the nucleus), respectively; n designates an uncharged neutron; and p designates a positively charged proton. All three of these processes resulted in unstable daughter nuclei that would subsequently disintegrate by β emission:

$$_Z^M A \to {}_{Z+1}^{M} A + \beta$$

This result was in fine accordance with the reigning theory of nuclear disintegration, which ruled out the emission of particles larger than the α particle (Figure 9). The categories in the contrast set for possible

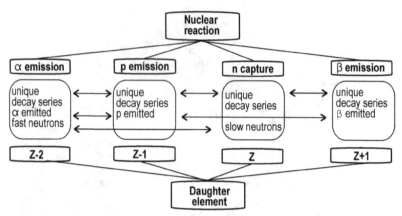

FIGURE 9. The contrast set of possible nuclear reactions after neutron bombardment. The kind hierarchy is linked to the kind hierarchy of daughter nuclei. Double-headed arrows indicate mutually exclusive properties: if a process instantiates any one property, then it cannot instantiate another property linked by double-headed arrows.

nuclear disintegrations are distinguished from each other by several features: daughter nuclei are different elements in the periodic tables and therefore have different chemical characteristics; different processes lead to distinct decay series with different half-lives; each process differs in the particles emitted. Further, through arguments based on the size of the potential barrier – the core idea of Gamow's treatment of decay – Meitner and Delbrück argued that the first two processes could not be produced by using slow neutrons. By the same token, they argued that the third process, neutron capture, was more likely for heavy nuclei, and that it was more likely for slow neutrons that contained little kinetic energy. These different characteristics were all expected to be correlated. If a process involved slow neutrons, it was expected that the nuclear reaction would be neutron capture, and that the nucleus produced would have the same chemical characteristics as the original nucleus. Alternatively, a daughter nucleus created by β emission would have the chemical characteristics of the next element ($Z + 1$) in the periodic table.

Having established the set of possible reactions in the case of artificially induced radioactivity, Fermi's team concentrated on heavy nuclei, especially uranium. Uranium, element 92, was the last known element in the periodic table. The question now arose whether the β decays involved in artificially induced radioactivity could lead to new elements with a higher atomic number than uranium. This question opened the possibility that the list of elements was not exhaustive; new elements that did not exist in nature could be produced artificially.

Fermi's team discovered five different reactions, which they distinguished by their half-lives. The product of one such reaction could be separated chemically from most heavy elements, and at the same time it seemed to have the chemical characteristics expected for element number 93. They concluded that they had discovered a new, transuranic element (Fermi 1934). This fell within the known contrast set of nuclear reactions for induced radioactivity, but it extended the contrast set of elements. However, this extension was an expected discovery. The contrast set of artificially induced disintegration processes indicated that elements with atomic numbers higher than that of uranium might very well exist, and it also provided the means to produce them as well as the classificatory means to identify them.

Although Kuhn himself gave few examples of kind hierarchies in the sciences themselves, our preliminary presentation of nuclear physics shows that the structures Kuhn described appear in real historical situations and operate in very much the manner he proposed. In this chapter we have also reviewed Kuhn's theory of concepts, in which contrast sets of objects are constituted through relations of both similarity and dissimilarity. This leads to an account of categorization in which objects falling under particular concepts bear no more than a family resemblance to one another, an idea introduced by Wittgenstein. We have pointed out the extent to which Kuhn goes beyond Wittgenstein in requiring empty perceptual space between sets of objects that can be grouped in this way, and we have considered the main application of Kuhn's theory to kind hierarchies. We have indicated that conceptual structures of this sort will also carry quasi-ontological knowledge and knowledge of regularities, and we have illustrated these features of Kuhn's account with a realistic historical example from the early history of nuclear physics. In the next chapter we will consider these same structures from the viewpoint of recent work on concepts in cognitive psychology.

3

Representing Concepts by Means of Dynamic Frames

A frame is a hierarchy of nodes (Figure 10). The origin of this notation may be traced to the British psychologist Sir Frederic Bartlett, who introduced the notion of a schema in his famous study of memory (Bartlett 1932). During the 1970s researchers in artificial intelligence developed and applied frames for a variety of purposes, including computer-based representations of everyday human activities (Schank 1975; Schank and Abelson 1977) and vision (Minsky 1975; Brewer 2000). During the 1980s, the American cognitive psychologist Lawrence W. Barsalou introduced frames in his studies of ad hoc categories (Barsalou 1982, 1991), autobiographical memories (Barsalou 1988), and contextual variability in concept representations (Barsalou 1987, 1989; Barsalou and Billmann 1989). He extended and refined previous presentations of the frame notation to represent concepts (Barsalou 1992b; Barsalou and Hale 1993), calling his new approach 'dynamic frames'. The present authors adopted his techniques in the 1990s and began to apply them to conceptual change in science and the implications of Kuhn's mature work (Andersen, Barker, and Chen 1996; Chen, Andersen, and Barker 1998; Barker, Chen, and Andersen 2003).

3.1 CONSTITUENTS OF DYNAMIC FRAMES

In the explanations that follow we will draw increasingly complex diagrams to represent frames, and we distinguish the concepts appearing

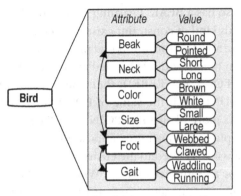

FIGURE 10. Partial frame for 'bird'.

in frames by capital letters when we mention them in the text. To represent a concept by a frame, one layer of nodes is selected to represent attributes of the concept. For the concept BIRD these include BEAK and FOOT. A second layer is selected to represent the many possible values of these attributes. In Figure 10, the central portion of the frame, enclosed here by a shaded box, shows these two layers of nodes as parallel columns, arranged vertically. The diagram is labeled a partial frame because there are many other attributes that might be included but are not listed here (nothing has been said about physical constitution, for example). There may be no single, definitive list that represents 'the' frame of the concept. However, all examples of BIRD share the properties in the attribute list such as BEAK, NECK, COLOR, (BODY) SIZE, FOOT, and GAIT. Attributes are listed in a particular order purely for convenience; the appearance of certain attributes at the top of the list does not indicate that they are more important than other attributes. Contrary to the conventional assumption that all features within a concept are structurally equal, the frame representation divides features into two different levels. A value is always attached to a particular attribute and every attribute must have a corresponding activated value in any given frame. Consequently, not all features within superordinate concepts are functionally equal.

Another kind of relation between nodes may also be recognized in frame diagrams. First, there are connections between nodes at the level of attributes. These connections are also called 'structural invariants'. Anything with a BEAK must also have a NECK, but not everything with a FOOT also has a BEAK. Learning a concept like BIRD involves learning that this kind of constraint exists between its attributes.

Second, there are also constraints that produce systematic variability in values, shown here by double-headed arrows. For birds in general, if the value of FOOT is WEBBED, then the value of BEAK is more likely to be ROUND, or if the value of FOOT is CLAWED, then the value of BEAK is more likely to be POINTED. These patterns may be understood as physical constraints imposed by nature. Webbed feet and round beaks are adapted to the environment in which water birds live but would be a hindrance on land. Wherever such constraint relations appear, knowing the value of one attribute (e.g., BEAK: ROUND) fixes the value of another (e.g., FOOT: WEBBED), and consequently limits the combinations of values available in the frame. Similarly, if the value of FOOT is WEBBED, then the value of GAIT is likely to be WADDLING. As indicated in Chapter 2.4, Kuhn discussed this form of conceptual relation, dubbed by Hoyningen-Huene "knowledge of regularities" (1993: 112–118). We will return to these issues in Sections 3.5 and 3.6.

Properties in the attribute and value lists, and the nodes representing superordinate concepts that lead to them, are said to be "activated" (by analogy with the selective activation of nodes in a neural network) when a particular subset is chosen to represent a specific subordinate concept. Conventionally, all of the attribute nodes are activated for every subordinate concept. However, value nodes appear in mutually exclusive clusters. Only one value for any given attribute may be activated, but different patterns of activation, or different choices of value, generate many different subordinate concepts, within the limits allowed by the attribute and value constraints already described. Each pattern of selection constitutes a subordinate concept; for example, a waterfowl is a bird whose values for BEAK and FOOT are restricted to ROUND and WEBBED.

A final important property of frames is that they are recursive. In principle any node in the frame may be expanded into a frame itself, which in turn contains nodes representing concepts, and so on (for further details, see the discussion of Figure 15). Where we make use of this property of the frames in later chapters we will draw a wide arrow leading from a node of one frame to a new frame showing attributes and values for the corresponding concept.

The recursive nature of frames deflects the seeming paradox that the frame, as a whole, represents a concept, but its elements, or nodes, are themselves concepts. This is not an atomistic form of analysis; there may be no ground floor or ultimate conceptual repertoire at which the chain of frames terminates. Similarly, there may be no single, unique way of drawing a frame for any given concept; there may be several equally defensible representations of a given concept or conceptual structure. A particular frame representation should be judged by its empirical adequacy as a representation of the behavior of a linguistic community, and beyond that it should be judged by its effectiveness as a problem-solving tool. Philosophers who expect the universe to divide into a single unique set of natural kinds may be displeased with this. However, to the extent that their belief in such natural kinds is alleged to be based on the actual use of language by nonphilosophers, including scientists, it is unfounded. There are no ultimate natural kinds in Wittgenstein's account of languages structured by games and family resemblance or in Kuhn's account of concept acquisition by learned relations of similarity and dissimilarity. Both accounts are vindicated by the empirical research in cognitive psychology we described in Chapter 1. This research led to the construction of the frame model we are now considering. The philosophers' theory of natural kinds therefore seems to be at variance with the way human beings actually use language and concepts. We will return to these questions briefly in Chapter 7. For present purposes, these issues can be deferred, until we have demonstrated the application of the frame representation to understanding conceptual change in the history of science.

The version of frame theory we adopt in this book takes the work of Lawrence Barsalou as its starting point. Barsalou's version of frames differs from most earlier ones in two main respects. First, he makes explicit use of differentiated layers to represent attribute-value sets. Unlike Minsky (1975), who put nodes at two different levels but only allowed an attribute to take a default value, Barsalou's frames allow an attribute to take several possible values. Second, Barsalou's frames allow the inclusion of constraints between some of the nodes in these layers, corresponding to the structural and attribute constraints already mentioned. From this point forward the reader is notified that whenever we speak of a frame, we mean a dynamic frame in Barsalou's sense.

Bird	Chair	Pants	Saw
feathers	legs	you wear it	make things
wings	seat	keep you warm	fix things
beak	back	legs	metal
legs	arms	buttons	handle
feet	comfortable	belt loops	teethblade
eyes	four legs	pockets	sharp
tail	wood	cloth	cuts
head	holds people	two legs	edge
claws			wooden handle
lays eggs			
nests			
chirps			
eats worms			
and flies			

FIGURE 11. Feature lists used in Rosch et al. 1976.

3.2 FRAMES IN HUMAN COGNITION

There are reasons to believe that frames are not merely a convenient notation that can elegantly describe the process of conceptual change. Recent cognitive studies indicate that frame diagrams accurately capture many important characteristics of human cognition. In this chapter, we suggest that frames should be understood as cognitive mechanisms that human beings use to obtain information from the environment, to store information in memory, and to retrieve information from memory. As cognitive mechanisms that reflect the information-processing capacities of our cognitive system, frames can offer explanatory models to account for conceptual change.

In concept representation, the primary alternatives to the frame approach have been 'feature-list models' like the prototype and exemplar theories described in Section 1.3.3. Work on natural categories in the 1970s typically represented concepts by lists of features that subjects produce, where a feature is any characteristic that the referents of the concept may process. Figure 11 shows some feature lists used in a study of basic-level concepts by Rosch and her collaborators (Rosch et al. 1976). For example, BIRD is represented by a list of features, including FEATHERS, WINGS, BEAK, FEET, and CLAWS. Since each of these features reflects a separate aspect of the concept's instances, it is regarded as independent and structurally equivalent to other features. Certain conceptual relations are implicit in the feature-listed model; for example, there is an *aspect* relation between each feature and the concept for which it is true, and there is an *and* relation connecting all the features of a concept conjunctively. However, the feature-list model does not specify any direct relations between features. For example, listing CLAWS as a feature of BIRD fails to represent the fact that claws are the shape of birds' feet. Moreover, it is very difficult, if not impossible, to illustrate the constraint relations between the values of BEAK and the values of FEET by using a feature-list representation (Barsalou and Hale 1993).

3.2.1. Evidence for Attribute-Value Sets

Cognitive studies have found that rather than treating all features as structurally equivalent we typically recognize certain hierarchical relations between features during categorization: that is, we know that some features are instances of others. Evidence for attribute-value sets is reported, for example, in people's understanding of stories (Stein 1992), in the process of combining concepts (Smith et al. 1988), and in the categorization of children's drawing (Wisniewski and Medin 1991). For our purposes, the most important evidence comes from studies of classification, where knowing the distinctions and connections between features at different levels of analysis is critical. For example, to know that a blueberry is an example of NONRED FRUIT, one must recognize that a certain set of values is related to a specific attribute. More specifically, one must be able to translate the negative category NONRED FRUIT into a positive one by recognizing both BLUE and NONRED as possible values of COLOR, and BLUE as a subset of NONRED. Many cognitive studies have suggested that people are able to distinguish between general and specific features and know that some specific features are values of the general ones. Accordingly, a central aspect of categorization is to identify those general features and to use them as the classification standards, because only they differ across categories (Barsalou and Hale 1993). For example, CAR cannot be distinguished from BIRD in terms of those features at a lower abstract level such as RED, HAVING WEIGHT, and CAN MOVE. Instead, CAR and BIRD must be classified by using attributes:

the former always has attributes ENGINE and TRANSMISSION while the latter has BEAK and WING.

More elaborate psychological experiments further confirm the existence of attribute-value sets and their importance in the process of categorization. These experiments usually involve two phases. In the initial, or learning, phase, subjects first receive paired stimuli, for example, a green circle paired with a red triangle and a green triangle paired with a red circle. They receive rewards for selecting the green circle or the red circle. In this phase the subjects are expected to learn that CIRCLE signals reward, but TRIANGLE does not, while COLOR is irrelevant.

In the second, or shift phase, subjects are again given paired stimuli, for example, a green circle with a red triangle. But they are now rewarded for selecting the red triangle, not the green circle. Finally, subjects are asked to identify the reward criterion used in the shift phase. In our example there are two possible answers: RED or TRIANGLE. If properties are processed as independent features, then there is no reason for subjects to favor one answer over the other. But if properties are actually processed as attribute-value pairs, then subjects who have learned that CIRCLE is rewarded in the learning phase should preferentially select the other value of the same attribute, TRIANGLE, as the reward criterion in the shift phase. Repeated experiments show that adults typically identify TRIANGLE as the reward criterion (Kendler and Kendler 1970). Subjects who learn that a particular attribute signals reward will continue attending to its values, even when these change (Barsalou 1992b: 26).

Experiments like these suggest that we encode stimuli as attribute-value pairs rather than as independent features, and that we typically pay more attention to attributes than to values. Further experiments show that the difference between interdimensional and extradimensional shifts increases with the developmental level. While adults and older children are able to separate attributes from values, young children typically do not differentiate them (Shepp 1978). The results of these experiments suggest that the ability to distinguish attributes from values reflects the developmental level of human cognition.

In addition to serving as classification standards, attributes function as generalizations in the process of category learning. Through a series of experiments, Ross, Perkins, and Tenpenny found that attributes are used by the subjects as generalizations that produce a reminding effect during the learning of a category (Ross, Perkins, and Tenpenny 1990). Their experiments include three phases. In the study phase, subjects learn about a small number of imaginary individuals and their features, such as an individual with features BUYS NAILS and LIKES ICE CREAM. Then in the first test phase, subjects are asked to categorize new imaginary individuals according to their similarities to the examples that they have learned in the study phase, such as an individual with features like BUYS WOOD and LIKES SHERBET. Ross and his collaborators found that, due to the similarities between the examples used in the study phase and the test phase, subjects place these two imaginary individuals in the same category and generalize that its members BUY CARPENTRY SUPPLIES and LIKE DESSERT. By functioning as attributes that can take different values, these generalizations affect the subsequent process of category learning. Subjects typically classify new examples to the established category if their features are values of the two attributes. For example, someone who BUYS A CHISEL belongs to the category, but someone who BUYS SUNGLASSES does not. These experiments again suggest that subjects do not represent the category by a group of features with a flat structure. Instead, they represent it with more abstract attributes that take other features as values. Because they are more abstract, attributes function as generalizations in category learning.

3.2.2. Evidence for Intraconceptual Relations

Cognitive studies have found that we typically do not treat features as mutually independent in the way that feature-list models suggest. Goldstone and his collaborators present a simple experiment as a prima facie counterexample to the assumption of feature independence (Goldstone, Medin, and Gentner 1991). In the experiment, subjects are first given three pairs of figures (Figure 12(a)) and asked to circle the pair that is more similar to the triangle pair. 89.3 percent of the subjects circle the pair of squares. Later, subjects are given three sets of figures (Figure 12(b)) and asked to circle the set that is more similar to the left-hand one. Of the subjects 100 percent circle the set with two circles and one square. From Figure 12a to Figure 12b the

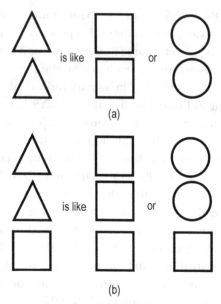

FIGURE 12. A counterexample of feature independence. Redrawn from Goldstone et al. 1991.

same feature, a single square, is added. If features are indeed independent of each other, adding the same feature to each set should not have reversed the original judgment of similarity. To explain the judgment reversal illustrated by this experiment, some kind of relation between features is required.

Cognitive studies have found that we typically have extensive knowledge about relations between features. Through analyzing natural categories such as BIRD, FURNITURE, FRUIT, FLOWER, TREE and CLOTHING, Malt and Smith found that people have a great deal of information about systematic property relations across the members of these categories. For example, people's knowledge of BIRD contains correlations between SING as a value of SOUND and SMALL as a value of SIZE, between LARGE as the value of SIZE and EAT FISH as the value of FOOD, and between WHITE/GRAY as the value of COLOR and NEAR SEA as the value of LOCATION (Malt and Smith 1984). These experiments demonstrate that features of natural categories are not randomly distributed, but rather are combined in a correlated fashion. Although some researchers assume that relations between features are simply correlations, there are reasons to believe that they are also conceptual. For example, although ROBIN and FEATHER are both correlated with BIRD, people know that a robin is a bird and that a feather is part of a bird. People would never claim that a feather is a bird or that a robin is a part of a bird (Barsalou 1992b).

Knowledge about relations between features affects categorization by modifying people's judgments of typicality. Medin, Altom, Edelson, and Freko found that people are sensitive to correlations during classification. They tended to give high typicality ratings to cases that preserve the correlations presented in the training phase, even when these cases contain fewer typical features (Medin et al. 1982). On the basis of their own experiments, Malt and Smith suggest that relations between features influence typicality judgments through certain particular perceptual or functional combinations. In fact, many of the feature correlations in our knowledge of BIRD are highly intercorrelated, for example, GRAY/WHITE and NEAR OCEAN, GRAY/WHITE and EAT FISH, and EAT FISH and NEAR OCEAN. These clusters of features give us the classification of BIRD, dividing it into groups such as water birds, songbirds, land birds, and, in this case, seabirds. The prototype of a category may be represented as a cluster of features belonging to one subset of the category members. Consequently, the typicality of an example decreases not only when it shares fewer features with the cluster, but also when it shares more features with a different, distinct cluster (Malt and Smith 1984).

Like the ability to distinguish attributes from values, the ability to recognize feature correlations reflects the developmental level of human cognition. Gentner presents evidence that our ability to recognize correlations increases with age. Specifically, Gentner found that children usually understand metaphors in terms of independent features while adults understand metaphors by relational structures (Gentner 1988). For example, the five-year-old children in Gentner's experiment typically described objects with adjectives or identified them by using concrete nouns, and they explained why "A cloud is like a sponge" by saying, "They both are soft" and "They both are fluffy." By contrast, the adults in the experiment typically explained metaphors in terms of relations expressed by transitive verbs ("X causes Y"), comparative adjectives ("X is longer than Y"), and prepositions ("X is inside Y"). They explained why "A cloud is like a sponge" by saying,

"They can both hold water" and "They can both give up water." Further studies indicate that our ability to recognize feature correlations is conditioned by time pressure. We tend to assume feature independence when time pressure increases. In other words, the more time subjects have to respond, the more likely they are to recognize feature correlations (Smith and Kemler-Nelson 1984). This suggests that the assumption of feature independence may be a simplifying tactic to deal with time pressure.

We have now described the frame model and some of the evidence for it. In the following sections we will show how the model can be used to represent the chief features of Kuhn's account of concepts.

3.3 FAMILY RESEMBLANCE AND GRADED STRUCTURE IN FRAMES

Two aspects of the frame account provide the flexibility required by Kuhn when he insists that any feature of an object can be used as the basis for grouping it into similarity classes with other objects. The first is the diversity of conceptual relations between a concept (BIRD) and its attributes. BEAK, COLOR, and GAIT are logically different kinds of attribute. Restricting the frame to concept-attribute links of only a single kind would dramatically limit the information represented and would fail to present the real complexity of the community's concept of BIRD. Kuhn's claim that any feature of an object can be used as a basis for grouping it into similarity classes corresponds to the claim that all kinds of attributes may appear in the same frame, regardless of logical type, and that any of them may be used in classifying objects. Second, any given individual need not employ the whole of a frame. The frame of a concept represents all the information connected with a given concept in a particular speech community. Individual speakers may know only part of this information but still use the concept correctly. Kuhn's claim that different individuals in the same speech community may use distinct features to classify the same objects corresponds to the claim that there is no restriction on which part of the community-defined frame must be employed by an individual.

Starting from the BIRD frame (Figure 10), to produce a frame for a particular type of bird, for example, a GOOSE (Figure 13), we

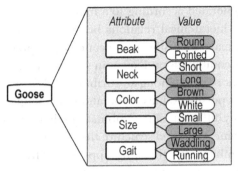

FIGURE 13. Partial frame for 'bird' with values for subordinate concept 'goose'.

distribute values across the attributes of the initial frame. However, our choices are not entirely free. As indicated in the previous sections, attributes and values are subject to a variety of constraints. For simplicity, let us consider only value constraints, as an example. Confining our attention to wild waterfowl, if we select LONG as the value for NECK, we are obliged to select LARGE for (BODY) SIZE. This is not a logical compulsion. Rather, our community's concepts of ducks, geese, and swans are such that these are the only choices we have.

The frame model of concepts accommodates graded structures, although skeptics in the research community continue to prefer versions of the traditional account of concepts. Earlier work by Barsalou and Sewell, using a version of prototype theory, cast considerable doubt on the idea that concepts are stable structures that are simply retrieved from long-term memory when needed. Instead, they suggested that concepts may be constructed instantly in working memory when they are needed (Barsalou and Sewell 1984: 36–46; Barsalou 1987). On the basis of results like these, other researchers concluded that graded structures primarily reflect differences in performance rather than in judgment. This encouraged them to believe that graded structures would disappear if we consider the essential information that makes up what they call 'conceptual cores', and they continued to prefer the traditional account of concepts in the face of the Roschian revolution (Rey 1985; cf. Armstrong, Gleitman, and Gleitman 1983). However, graded structures may still exist when essential information from conceptual cores is taken into consideration. In the frame model, a concept is generated according to a fundamental representational structure. The hierarchical relations between attributes and values within a frame make it possible that a concept is represented by many different value combinations. Except in some extreme cases in which strong connections exist among all attributes and their values, a frame seldom reduces possible value combinations to a single set. A concept usually has many different value combinations that may be represented by the same frame. A graded structure thus emerges naturally among the possible combinations: those that appear most frequently (or those that are best suited to achieve the goal served by the concept) become the prototype (Barsalou 1992a, 1993).

By using frames it is easy to construct a diagram showing how similarity classes may be constructed by using similarity and dissimilarity relations that have no single common feature. This is the mark of a family resemblance concept and shows the compatibility of the frame account with family resemblance as a general feature of concepts.

Let us return to the partial frame for BIRD in Figure 10 and use it to generate a partial frame for several particular kinds of birds, say ducks, geese, and swans. Recalling Kuhn's example of the child learning these categories, we will assume that they are wild ducks, geese, and swans, of the kind Johnny is likely to see at the park. For purposes of this illustration, we will indicate activated nodes by shading. A goose is a bird with a large body that is brown in color, a rounded beak, a long neck, and a waddling gait. Thus, in Figure 14, we indicate by shading which specific nodes have been activated for the attributes listed. Only a selection of the value nodes are activated. Compared to geese, ducks have small bodies, short necks, and rounded beaks. Swans, however, have similar size bodies, are white rather than brown, and have rounded beaks similar to ducks.

The highly simplified and schematic rendering of the concepts in Figure 14 is already sufficient to illustrate that BIRD is a family resemblance concept. The three subconcepts, DUCK, GOOSE, and SWAN, have no single common feature (activated value) except GAIT and BEAK. Any bird that lacks webbed feet or has a pointed beak (for example, a thrush or an American robin) may be included in the concept BIRD without sharing a single common feature with the birds already introduced. This shows the compatibility of the frame account

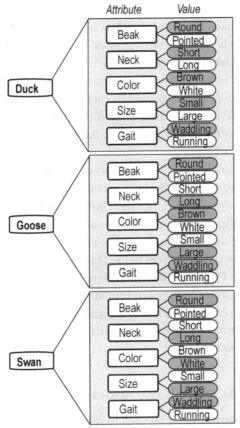

FIGURE 14. Partial frames for 'duck', 'goose', and 'swan' with activated values indicated by shading.

with the family resemblance account. However, it should be added that the expectation in both accounts is that adding further subconcepts will generate a structure like the one we have described, with no single common feature among the activated values, in contrast to the traditional theory, which would expect to reveal either a set of necessary and sufficient conditions or a conceptual core common to all subconcepts.

What Kuhn calls similarity and dissimilarity relations are not depicted directly in frame diagrams. Rather they are implicit in the recognition of a contrast between the different values that a given attribute can take. Recognizing similarities and dissimilarities between objects on the basis of the values taken by specific attributes is the way we group objects into similarity classes (DUCK, GOOSE, etc.). In a frame representation, this comparison is made by inspecting the activated attributes in frames for concepts or subconcepts at the same level (for example, the values activated in each of the three frames making up Figure 14). Two groups are similar if they have the same value for a particular attribute, and they are dissimilar if they have different values for the same attribute.

Thus, the frame account of concepts naturally accommodates family resemblance, as well as Kuhn's account of concept acquisition in terms of similarity and dissimilarity relations, while not requiring that concepts be defined through necessary and sufficient conditions. The same resources that support the family resemblance account also permit overt or covert divergences in the features used by a community to classify objects. The frame model is therefore not unique in its ability to capture important features of human conceptual systems revealed by empirical research and historical investigation, but it has a number of advantages, one of which is its ability to represent kind hierarchies, to which we now turn.

3.4 FRAMES AND KIND HIERARCHIES

A frame like Figure 10 may be used to represent the kind hierarchy of birds. It indicates that there is an inclusive relation between the superordinate concept BIRD and the subordinate concepts WATERFOWL and LAND BIRD and it also indicates the contrastive relations among concepts within the same subordinate group, because WATERFOWL and LAND BIRD should never be applied to the same object. It is acceptable to call a waterfowl a bird because the concept of the former is subordinated to the concept of the latter in the frame, but not to call it a land bird. In other words, subordinate concepts of the same superordinate concept cannot overlap in their referents, and so no object is both a waterfowl and a land bird. We shall discuss both the inclusive relation and the contrastive relation in more detail in Section 4.2. Here we note that in the frame representation, both the inclusive and the contrastive relations are embedded in the internal structure of the superordinate concept. The inclusive relation derives from the attribute list: all subordinate concepts belong to the superordinate one because their instances share the properties listed as attributes for the superordinate. The contrastive relations derive from the pattern of the activated values: two subordinate concepts contrast if they have different values for the same attribute.

Frames and kind hierarchies are different. Kind hierarchies are built from only a single type of node, representing entities. The relations between nodes are also limited to the relation of instantiation or class inclusion. Any such hierarchy can be converted to a frame by inserting the intervening layers of nodes corresponding to the attributes and values of the superordinate concept. However, frames, by contrast, admit very different sorts of concepts among nodes at the same level and admit many different relations between nodes. Although most of the attribute nodes in Figure 10 happen to be body parts, several other attributes, and many others that might legitimately appear in the frame, are not. The color and body size of a bird are equally attributes and may function in classifying the bird, but they are not parts of its body, in the usual sense of 'part' (detachable subunit). Similarly the gait of a bird may be a useful classification standard for distinguishing waterfowl that waddle from land birds that run. 'Gait' is clearly not an attribute in the same class as either body parts, or shape, or color. An additional important difference between frames and kind hierarchies is the recursive nature of frames. In principle any node in a frame (including the individual nodes making up the attribute and value layers) can be expanded into a new frame. For example, consider again the frame for BIRD (Figure 10). All the attributes are concepts that can themselves be represented by a frame. A partial frame for FOOT might include the attributes JOINTS and SKIN. Hence, the frame for BIRD can be expanded by combining the two frames (Figure 15). Frames can also be expanded to accommodate a kind hierarchy with several levels. For example, consider what we would need to include in the frame for ANIMAL (Figure 16). A partial frame for ANIMAL might have the attributes LIMBS, RESPIRATORY ORGAN, and METHOD OF REPRODUCTION. Different value distributions for these attributes result in the three subordinate concepts BIRD, FISH, and MAMMAL. Each of these concepts may again be represented by a frame, and value distributions in this frame result in new sub-subordinate concepts, for example,

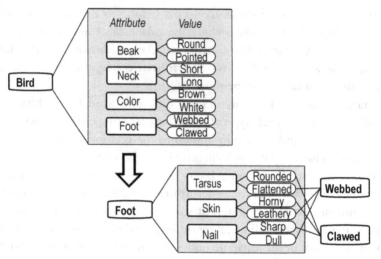

FIGURE 15. Frames are recursive. Here the attribute 'foot' in the frame for 'bird' is expanded into a partial frame with its own attributes and values.

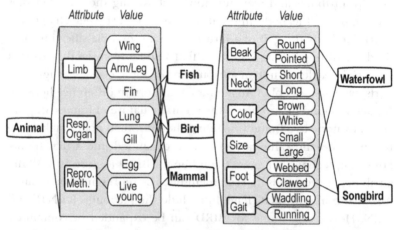

FIGURE 16. A frame representation of a multiple-level kind hierarchy.

WATERFOWL and SONGBIRDS as subordinates to the subordinate concept BIRD.

In earlier frame diagrams activated value nodes were indicated by shading. Here the activated values are indicated by lines linking the node for FISH to the nodes for FIN, GILL, and EGG. Other lines link different values to the nodes for BIRD and MAMMAL and, at the extreme right, WATERFOWL and SONGBIRD. This enables us to display the contrasting value sets for several subordinate concepts in a single frame diagram for their superordinate concept.

Such an extended frame also displays the inheritability of specific values for attributes of superordinate kinds, that is, values that hold for all instances from a given level in the kind hierarchy and also hold for their subordinates. For example, the value WING of the attribute LIMB holds for all subordinates to the concept BIRD, while the value WEBBED of the attribute FOOT holds only for a particular subclass of birds. This aspect of frames will be important when discussing degrees of incommensurability (Sections 5.3 and 5.4). In contrast to frames, kind hierarchies are not recursive. Although intermediate concepts in a kind hierarchy may themselves have taxonomic structure, the lowest level in the kind hierarchy consists of entities that do not admit of further classification. Because of these differences, the frame notation may be used to represent concepts and structures of concepts that cannot be represented by kind hierarchies.

3.5 KNOWLEDGE OF REGULARITIES AND ONTOLOGICAL KNOWLEDGE

Unlike kind hierarchies, frame representations also display a possible cognitive mechanism behind the classification process. The frame of a superordinate concept directly determines the possible concepts at the subordinate level. For example, since the frame of BIRD in Figure 10 has 6 attributes and each of them has 2 possible values, there are 64 possible property combinations (2^6) and thereby 64 possible concepts at the subordinate level. (As the frame is only partial, this is a minimum number.) However, value constraints exclude some of these combinations, either because these combinations are not found in nature or because they are excluded for theoretical reasons (we shall return to this in Section 3.6). If this frame is adopted, then there are no instances of BIRD with BEAK: ROUND and FOOT: CLAWED, or with BEAK: POINTED and FOOT: WEBBED. The results are only two property combinations (BEAK: ROUND with FOOT: WEBBED and BEAK: POINTED with FOOT: CLAWED), which form two subordinate concepts – WATERFOWL and LAND BIRD. In this way, the frame specifies classification standards: birds are classified according to their beak and foot. Although this example is too simple, it may correspond to one of the historical stages in the development of classifications for birds. We present a more realistic historical case in Sections 4.3 and 4.5.

As we saw in Section 2.4, an important implication of Kuhn's theory of concepts is the close relation between knowledge of regularities and ontological knowledge. We are now in a position to show how the frame representation displays a possible mechanism behind these two aspects of knowledge contained in concepts and explains why they cannot be separated.

We just saw that the frame determines which potential concepts are possible at the subordinate level but that constraints exclude some of these. If we examine the frames for waterfowl developed by Johnny after his encounter with ducks, geese, and swans (see Section 2.2), Johnny possesses three subordinate concepts: DUCK, GOOSE, and SWAN (Figure 14). Value constraints determine that these are the only possible subordinates (although, if Johnny had met different birds, the original frame might be used to generate subordinate concepts for other waterfowl). However, the value constraints also determine that only specific activation patterns are possible. Within a given activation pattern all activated values are equal. There is no distinction between values that must be used to identify instances of a concept and values that show how an already identified instance will behave. So the activation pattern contains at the same time both ontological knowledge and knowledge of regularities. For example, in the frame for the concept DUCK, all values in the activation pattern BEAK: ROUNDED, NECK: SHORT, COLOR: BROWN, SIZE: SMALL, and GAIT: WADDLING are equal. Johnny may identify ducks by their small size and brown color, knowing that ducks have short necks, or he may identify ducks instead by their short neck and rounded beak, knowing that ducks have a small body.

3.6 VALUE CONSTRAINTS AND CAUSAL THEORIES

Value constraints play an important role in embodying both ontological knowledge and knowledge of regularities. However, it has been argued by cognitive psychologists that

[f]eatures in categories are not correlated by virtue of random combinations. Rather, correlations arise from logical and biological necessity: Animals and

artifacts have structural properties in order to fulfill various functions, so that some structural properties tend to occur with others, and certain structures occur with certain functions. It is no accident that animals with wings often fly or that objects with walls tend to have roofs. Even less obvious correlations, such as the one between furniture being made of wood and also having a flat top . . . , usually have clear explanations. (Murphy and Medin 1985/1999: 439)

The connections Murphy and Medin postulated are best understood, for our purposes, as causal connections. What they claimed is that people tend to deduce causal explanations for attribute correlations, a view that has become known as the Theory-Theory of concepts (Margolis and Laurence 1999: 43–51). Thus, they believed that "feature correlations are partly supplied by people's theories and that the causal mechanism contained in theories are the means by which correlational structure is represented" (Murphy and Medin 1985/1999: 431). On this view, value constraints are not simply empirical generalizations expressing patterns of values that have been discovered accidentally. Instead, theories are needed to *explain* the value constraints. For example, in the case of waterfowl, the correlation between the value WADDLING of the attribute GAIT and the value WEBBED of the attribute FEET is not just an accident, but may be explained by the use of the feet to produce the gait. In the frame representation, this causal connection is expressed as an attribute constraint between the attribute FEET and the attribute GAIT that specifies how the former attribute is involved in the production of the latter (Figure 10).

Versions of the Theory-Theory have been at the core of much work on conceptual development in childhood (e.g., Carey 1985, 1991/1999; Keil 1989). In the 1990s, this approach was further refined by detailed studies of the role of causal status in determining the centrality of individual attributes (e.g., Ahn 1998; Ahn and Dennis 2001; Ahn et al. 1995; Sloman, Love, and Ahn 1998).

There is considerable divergence in the literature on the goals and characteristics of the Theory-Theory, and on what such a 'theory' might be (for an overview see Margolis and Laurence 1999). Contrary to our aim in this book, some people hope that the 'theories' introduced in the Theory-Theory will restore essentialism or at least "respect people's tendency towards essentialist thinking"

(Margolis and Laurence 1999: 47). However, there is no overall agreement among theory-theorists on the issue of essentialism. Although the focus on underlying causal explanations of the correlation of surface attributes might seem to encourage essentialist views, Murphy and Medin have emphasized that the features that appear essential are not so because of the structure of the world, but because they are the features that are most central to our current understanding of the world (see Murphy and Medin 1985/1999: 454). Later versions of the Theory-Theory, such as the causal status interpretation of attribute centrality developed by Ahn, also point out that people tend to give weight to features that are seen as the causes of other features more than to effects (see Ahn 1998: 138). However, Ahn also makes clear that this view is different from essentialism in that causal features need not be defining features and, by the same token, that features need not be dichotomized into essential and nonessential features. To illustrate this Ahn points out that a feature may be the cause of one feature and at the same time the effect of another, adding that "although it might be possible to conjecture that the most terminal cause (which is an essence in the essentialist framework) serves as a defining feature, the causal status hypothesis in its current form is mute about the debate on whether or not concepts have (or are believed to have) defining features" (Ahn 1998: 163).

Dissociating the Theory-Theory from essentialism has important implications for conceptual change. On Murphy and Medin's account, a discovery of mismatch between attribute correlations and their underlying explanation may lead to fundamental change in a conceptual structure: "If it turned out that *carrots* weren't made of cells, then we would have to reconsider most of our other beliefs about carrots as well as about plants in general (for example, our theories of plant growth)" (Murphy and Medin 1985/1999: 452, italics in original). This is developed further in the work of Ahn, who points out that

[t]he more causal a feature is, the more difficult it seems to mutate the feature without changing other aspects of the conceptual representation. For instance, if we are to imagine a new breed of dog that does not have a hippocampus (which presumably causes many behaviors of dogs), we need to alter a lot of features in our dog concept including their behavior and even their status as

a pet. On the other hand, imagining a new kind of dog that does not have a tail would not require much conceptual mutation except that they would not wag their tails. As such, more causally central features seem more responsible for conceptual coherence and consequently would be judged more central in categorization. (Ahn 1998: 140)

In Chapter 4 we shall return to the mismatch between feature correlations and underlying explanations as one of the mechanisms responsible for conceptual change, in a further discussion of the discovery of nuclear fission.

Perhaps as a result of the lingering appeal of essentialism, some writers treat the concept of theory as primary and treat concepts in general as entities to be explicated *after* the concept of theory has been introduced (Gopnik and Meltzoff 1997, cited in Margolis and Laurence 1999: 51). It should be clear that we are approaching matters in the reverse of this order: frames give a general account of concepts that includes attributes and values; only some of these attributes and values may be linked by constraints, and only some of these constraints may be the kind studied by the Theory-Theory. Among them we single out for special attention those that correspond to claims of causal connection. (And to prevent further difficulties with the concept of 'cause' we will take as a working hypothesis that the relevant sense of 'cause' is the one understood by the historical actors we are studying in a particular episode of scientific change.) We believe that constraints between attributes and values play an important role in some of the conceptual structures we discuss, and that it is important to acknowledge the possible connection between these constraints and the 'theories' postulated in the Theory-Theory. The Theory-Theory is therefore a supplement, not a replacement, for frame theory.

We have now introduced all the main features of the frame theory of concepts. We have highlighted the empirical evidence that supports the differentiation of attributes and values within conceptual structures and discussed the way in which subordinate concepts can be generated from the frame for a superordinate and how this process lends itself to the representation of family resemblance and graded structure. We have also discussed the connection between this account and Kuhn's categories of ontological knowledge and knowledge of regularities. Last, we have suggested that some constraints between attributes and values may be equivalent to theories, or, more simply, causal connections. In our view all these structures are revisable in the course of the historical development of science. In the next three chapters we apply these ideas to retrieve and extend the most important features of Kuhn's model for the development of science.

4

Scientific Change

.

In Chapter 2 we gave an account of concepts and conceptual structures based on family resemblance. We showed that on this account possession of a conceptual structure implies knowledge of ontology, as objects not belonging to any of the known similarity classes are assumed not to exist. Likewise, we showed that through the relations of similarity and dissimilarity, possession of a conceptual structure implies knowledge of regularities, that is, expectations of the different situations that nature does and does not present.

In Chapter 3 we have seen how conceptual structures of the kind introduced by Kuhn may be represented by dynamic frames, a form of representation developed in cognitive psychology and independently supported by empirical research. Frames not only accommodate the most important features of Kuhn's account, such as family resemblance, but may also be used to represent graded structure, the most important empirical phenomenon documented by studies of categorization supporting the reality of family-resemblance categories. The frame account allows us to display details of conceptual structures that are otherwise difficult to examine, such as the patterns of attribute-value sets that characterize concepts, and it allows us to locate constraints between elements of the structure that correspond to knowledge of ontology and knowledge of regularities.

We have already suggested a developmental perspective: a particular conceptual structure is always given by the preceding generation, which passes it on to the next. In this way, new generations are continuously socialized into the language community that exists at any particular time, and in the process of this socialization they inherit the current knowledge of ontology and the knowledge of regularities implied by the existing conceptual structure.

However, the next step in the account of concepts and conceptual structures is to explain *change*. Further, the account of conceptual change must explain how such changes affect the knowledge of ontology and knowledge of regularities implicit in the conceptual structure. In this way our account of conceptual change may explain the development of science through the development of underlying conceptual structures and clarify the notion of incommensurability as a relation that holds between a conceptual structure and its historical successor.

4.1 THE PHASE MODEL OF SCIENTIFIC DEVELOPMENT

In *The Structure of Scientific Revolutions* Kuhn claimed to have described, in a preliminary way, a pattern of development that could be found throughout science and throughout science's history. He used historical examples ranging from ancient astronomy and optics to physics in the twentieth century. But from the viewpoint of the historian, it is dramatically implausible to suggest that the usual factors considered in a historical explanation were sufficiently constant over all the periods considered by Kuhn to yield similar structures in each one. Between the ancient period and the twentieth century, the institutional structure of science, its relations to the wider culture, and the education, social class, and career paths of scientists themselves changed not once but several times. Despite his insistence that the scientific community is the main actor in his account, Kuhn was adamant that such factors played little role in the intellectual changes that were his primary concern.

Rejecting the usual historical factors, a second possibility to justify the appearance of similar structures in different disciplines and different periods might be the cognitive structures that have now been demonstrated by psychologists and cognitive scientists to be universal features of human intellectual activity. On this basis, the division between normal and revolutionary science can be understood as the distinction between research conducted in terms of an existing conceptual structure without changing that structure, and research proceeding by modifying an existing conceptual structure.

In principle, we should not see this division as corresponding to a linear sequence of historical changes, with normal science succeeded by revolutionary science, succeeded by normal science, indefinitely. Both patterns of research may coexist. Likewise, we do not suggest that this division applies only to revolutions that involve major modifications of conceptual structures. Instead, the division implies that revolutions may vary in scope and severity from minirevolutions created by modifications of minor parts of the conceptual structure to major revolutions created by fundamental modifications affecting large parts of the conceptual structure. However, we shall also suggest reasons for the conservative nature of normal science, and for the relative infrequency of the extensive changes in conceptual structures that we recognize as major revolutions.

4.2 HIERARCHICAL PRINCIPLES OF STABLE CONCEPTUAL STRUCTURES

As explained in Section 2.3, concepts formed by family resemblance are tied together in contrast sets that form kind hierarchies. The formation of kind hierarchies by breaking up a class into subclasses has been an established part of logic since antiquity. Usually, this logical division is characterized by three principles: a principle of no-overlap, a principle of exhaustion, and a principle of inclusion. Although we will subsequently express reservations about the extent to which these principles are distinguishable or logically independent when applied to actual historical cases, some version of these principles has been taken as fundamental for any hierarchy of kinds, and violations of the principles therefore indicate that something is wrong with a hierarchical structure.

4.2.1. The No-Overlap Principle

According to the no-overlap principle the division in a kind hierarchy is exclusive: no concepts in a contrast set formed by division of a superordinate are allowed to overlap. The periodic table of the elements is such an exclusive division: no atom can be both iron and carbon. Likewise, for astronomers in the sixteenth century, there are no fixed stars that are also planets, nor celestial objects that are also sublunar objects.

The no-overlap principle is violated by objects that cannot be classified in this way. If an object is encountered, that judged from some
features has to be an instance of one concept, but judged from other
features has to be an instance of another concept in the same contrast
set, that object is an anomaly. We shall return to historical examples
of this sort in Section 4.3.

4.2.2. The Exhaustion Principle

According to the exhaustion principle a division of a superordinate
concept never leaves any residual instances: that is, the extensions of
all concepts in a contrast set together exhaust the extension of their
superordinate.

From a chemistry textbook we can learn the full contrast set of all
naturally occurring elements. This contrast set will be exclusive in the
sense that we do not expect to find an element that does not belong
to any of the existing ninety-two categories from hydrogen to uranium
but still occurs naturally. Likewise, astronomers in the sixteenth century knew that celestial objects were either fixed stars or planets with
individual motions (a category that included the sun and the moon).
They did not expect to find anything that did not belong to one of
these categories but could still be counted as a celestial object.

This principle can be violated if an object is encountered that
judged from some features clearly belongs to a given contrast set, but
that judged from features usually differentiating the contrasting concepts in the set cannot be an instance of any of these concepts.

4.2.3. The Inclusion Principle

According to the inclusion principle, all instances of a subordinate concept are also instances of the superordinate concept. In other words,
if a superordinate concept in a kind hierarchy has certain properties,
then any object that belongs to a subordinate level should also have
these properties; if all birds lay eggs, then so do waterfowl and so do
ducks. Conversely, this principle excludes waterfowl that are not birds.
Similarly, sixteenth-century astronomers were certain that they would
find no stars that were not celestial objects.

This principle can be violated if the object exhibits features according to which an object that, due to some features, clearly is an instance of a specific concept either (1) could at the same time be an instance of a concept contrasting with its superordinate or (2) cannot be assigned to any of the concepts in the superordinate contrast set. Consider, for example, the claim "The earth is a planet" from the viewpoint of a sixteenth-century natural philosopher who accepts an exclusive distinction between celestial and terrestrial objects. Followers of Aristotle and Ptolemy accepted planets as one category of celestial object. Claiming that the earth is a planet ostensibly places it in the same category as other planets, making it a celestial object. But the superordinate categories CELESTIAL and TERRESTRIAL form a contrast set. The earth is by definition terrestrial. So the claim "The earth is a planet" either locates the earth among the contrast set to TERRESTRIAL or suggests that the earth cannot be assigned to any of the concepts in the superordinate contrast set TERRESTRIAL-CELESTIAL. In the former case the situation can also be described as a violation of the no-overlap principle on the superordinate level, in the latter case as a violation of the exhaustion principle on the superordinate level. Thus, violations of the inclusion principle may turn out to be violations of one of the two earlier principles rather than a separate category.

4.3 ANOMALIES AS VIOLATIONS OF THE HIERARCHICAL PRINCIPLES

On the basis of the characteristics of stable conceptual structures expressed through the three hierarchical principles, we are now in a position to analyze the notion of anomalies in more detail.

On our account, anomalies are findings that run counter to our expectations about what exists in the world and which characteristics these objects and phenomena have. But findings that run counter to these expectations cannot be made easily. It is difficult to discover a phenomenon or an object that was never anticipated to exist, since there is simply no category by which to classify it. Only after this category has been formed can the anomaly be recognized as an actual phenomenon or object. Until then it is perceived simply as 'something wrong'.

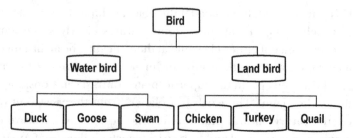

FIGURE 17. Kind hierarchy showing the contrast set of water birds and land birds.

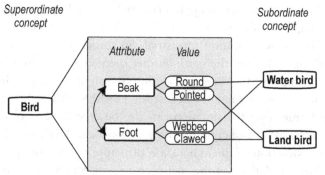

FIGURE 18. A partial frame for the concept 'bird' with the two subordinates 'water bird' and 'land bird'.

Returning to the characteristics of stable conceptual structures we can illuminate when anomalies are perceived as 'something wrong'. The hierarchical principles must all be obeyed for any kind hierarchy, and objects that violate any of the principles are therefore at first simply 'something wrong'. As explained previously, violations of the hierarchical principles arise when an object is encountered that judged from different features will be categorized into different contrasting categories. Hence, the first recognition of an anomaly as simply 'something wrong' can, at least for normic concepts, be analyzed in terms of violations of the hierarchical principles.

For example, we can imagine a simple kind hierarchy of birds that includes the two subordinate concepts WATER BIRD and LAND BIRD. This simple kind hierarchy (Figure 17) can be represented by a frame (Figure 18) with the two attributes FOOT and BEAK. Water birds have webbed feet and rounded beaks, whereas land birds have clawed feet

FIGURE 19. Horned screamer (*Anhima cornuta*).

and pointed beaks. If we now encounter a bird like the South American screamer that has webbed feet and a pointed beak (Figure 19), this instance will violate the no-overlap and the exhaustion principles. Judged on its feet it ought to be a member of the category WATER BIRD, but judged on its beak it ought to be a member of the category LAND BIRD. By the same token, we see in the frame representation that the activated pattern of values that represents SCREAMER does not correspond to a single subordinate concept but to a (simple) mismatch of the two contrasting categories (Figure 20).

In their initial phase, anomalies are perceived primarily as 'something wrong'. Much recent work on anomalies has focused on the next step in the process, namely, the cognitive processes of anomaly *resolution*. Darden has described the kind of scientific reasoning involved not only in localizing an anomaly but also in generating new hypotheses that can account for it or dissolve it (e.g., Darden 1992, 1998). In some cases anomalies may be explained away without requiring much change, for example, by claiming that the anomalous case is simply not a normal instance, but a monstrous one. This is what Darden calls a "monster anomaly" (Darden 1992: 258ff., 1998: 142). Other anomalies are accommodated by changes in conceptual structures that permit the

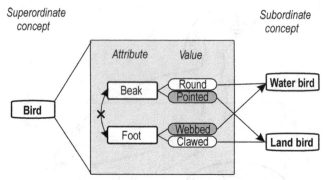

FIGURE 20. Violation of the hierarchical principles shown in the frame representation.

phenomenon to be viewed instead as instances that are no longer monstrous but normal. These Darden calls "model anomalies" because they serve as models of the normal types of processes that are commonly found (Darden 1992: 259, 1998: 143). We shall now examine a simple example showing how an anomalous instance may serve as a model of the normal and lead to the introduction of new categories.

In a simplified way, the hypothetical example about overlap between the contrasting concepts WATER BIRD and LAND BIRD developed earlier mirrors the development of ornithological kind hierarchies during the Darwinian revolution. An examination of the sequence of historical changes that occurred in ornithology during the nineteenth century will therefore illustrate some of the mechanisms of conceptual change triggered by model anomalies. To illustrate the generality of this kind of analysis we shall also return to the historical case study about the earliest research on transuranic elements that we introduced in Chapter 2. This will allow us to consider several different kinds of revision in conceptual structures that can occur in response to anomalies and explain the differences between revolutionary and nonrevolutionary developments.

4.3.1. Sundevall's Taxonomy: Conceptual Revision in Normal Science

In the seventeenth century when the first ornithological taxonomy was developed (Ray 1678), birds were simply divided into two classes,

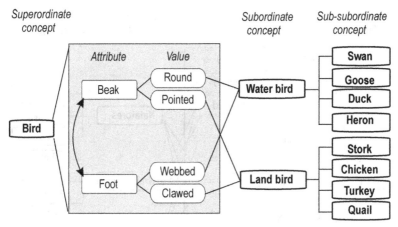

FIGURE 21. A partial frame representation of the Ray taxonomy.

'water bird' and 'land bird', according to their beak shape and foot structure (Figure 21). Typical examples of 'water bird' were those with a round beak and webbed feet like ducks or swans, and typical examples of 'land bird' were those with a pointed beak and clawed feet like chickens or quail. By the early nineteenth century, however, many newly found birds could not be fitted into this two-category system. As already mentioned, a South American bird called a screamer was found to have webbed feet like a duck but a pointed beak like a chicken (see Figure 19).

To accommodate such anomalies a popular taxonomy proposed by Sundevall in the 1830s adopted attributes, including BEAK SHAPE, PLUMAGE PATTERN, WING-FEATHER ARRANGEMENT, LEG FORM, and FOOT STRUCTURE, as classification standards (Sundevall 1889). The five attributes generate more allowed property combinations, and thereby more possible concepts (Figure 22). In this way, Sundevall converted the anomalous instance into an example of the normal by incorporating a new category in the taxonomy. The Sundevall taxonomy was more flexible than the old two-category system and was able to accommodate birds like the screamer that were anomalies in the old system. Because BEAK and FOOT are no longer related in the Sundevall system, it becomes possible to have a property combination that includes both BEAK: POINTED and FOOT: WEBBED, the key features of screamers. In this way, Sundevall eliminated the

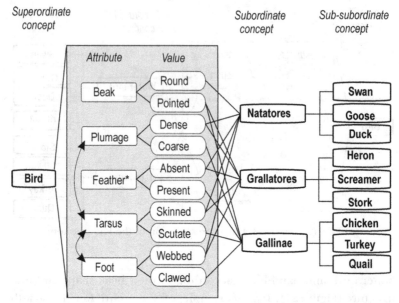

FIGURE 22. A partial frame for Sundevall's concept 'bird' and its taxonomy.
* The attribute 'feather' refers to the existence of the fifth secondary.

anomaly by putting SCREAMER under a new category GRALLA-
TORES, independent of WATER BIRD and LAND BIRD, which
are now replaced by the categories NATATORES (swimmers) and
GALLINAE (chicken-like), respectively. What were previously anomalies had now become instances of the normal.

Although Sundevall's taxonomy causes a redistribution of referents,
there are crucial continuities with Ray's taxonomy. Many dissimilarity
relations from the old taxonomy are preserved after the taxonomic
change. For example, the dissimilarity relations between WATER BIRD
and LAND BIRD in the old taxonomy also exist in the new one, where
NATATORES and GALLINAE continue to have opposite values in
BEAK and FOOT. Also, the newly added dissimilarity relations do not
contradict the preserved ones. The new dissimilarity relations between
NATATORES and GRALLATORES, deriving from the opposite value
assignments in the attributes of PLUMAGE, FEATHER, and LEG, do
not alter the dissimilarity relations inherited from the old taxonomy.
Sundevall's introduction of the new category GRALLATORES is a typical example of anomaly resolution in normal science. The change

from Ray's taxonomy to Sundevall's might be presented as an instance of the weakest form of conceptual change.

To display in greater detail both similar processes in which anomalies serve as model anomalies in the creation of new categories and processes in which anomalies are treated as monster anomalies, we return to the history of nuclear physics during the period preceding the discovery of fission.

4.3.2. Core Concepts of Nuclear Physics in the 1930s

In Chapter 2 we described the development of the core categories used in understanding radioactivity up to the early 1930s. We shall now analyze how research on induced radioactivity led to the discovery of nuclear fission. This research focused on the bombardment of heavy nuclei with neutrons. As we explained in Chapter 2, this could lead to α emission, proton emission, or neutron capture, which would all result in unstable daughter nuclei that would subsequently decay, usually by β emission. We showed that several features were used to distinguish the contrasting categories: the daughter nuclei were different elements in the periodic table and therefore had different chemical characteristics; the daughter nuclei decayed with particular half-life periods; different particles were emitted in the different processes; and some of the processes could only be produced by neutrons with characteristically high, or characteristically low, energy. These different characteristics were expected to correlate in specific ways, and theoretical models like Gamow's model of α decay explained several of the correlations.

We shall now examine the concepts involved in induced radioactivity using the frame model. As explained in Chapter 3, a concept is represented by two layers of nodes: one layer representing its attributes and a second layer representing the many possible values of these attributes. Further, we explained how specific relations between attributes and relations between specific values both determine the range of concepts that may be instantiated through the frame (what we called quasi-ontological knowledge) and represent the knowledge of regularities contained in the concept.

We will consider experiments in which the last known element in the periodic table, uranium, is bombarded with neutrons. In this case, a

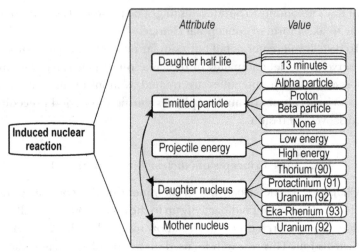

FIGURE 23. A partial frame for 'induced nuclear reaction' in the case of uranium bombarded with neutrons.

partial frame for induced nuclear reactions may, for example, have the following attributes: half-life of the daughter nucleus, emitted particle in the primary process, projectile energy, daughter nucleus, and mother nucleus. Each of the attributes can have different values. For some attributes the possible values are restricted to a small set of well-defined values, like the attribute EMITTED PARTICLE that can have the values α PARTICLE, PROTON, β PARTICLE, or NONE. For other attributes the possible values are drawn from a large set of values, like the attribute DAUGHTER HALF-LIFE (Figure 23). The major structural invariant in this frame is the attribute relation among EMITTED PARTICLE, MOTHER NUCLEUS, and DAUGHTER NUCLEUS since the daughter nucleus is produced by the emission of the particle from the mother nucleus. This structural invariant determines the patterns of value distributions among the three attributes according to the three reaction schemes:

$$_Z^M A + {}^1n \rightarrow {}_{Z-2}^{M-3} A + \alpha$$

$$_Z^M A + {}^1n \rightarrow {}_{Z-1}^{M} A + p$$

$$_Z^M A + {}^1n \rightarrow {}_{Z}^{M+1} A$$

where Z is the proton number and M the mass number of the nucleus, for an element A. The reaction scheme provides an underlying

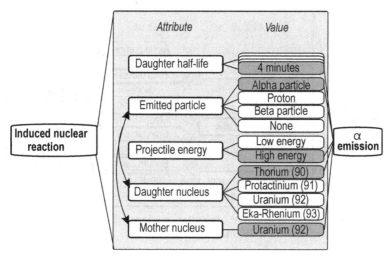

FIGURE 24. Partial frame for 'induced nuclear reaction'; the subordinate concept 'α emission' is instantiated.

explanation of value constraints among values of EMITTED PARTICLE, MOTHER NUCLEUS, and DAUGHTER NUCLEUS. Other important value constraints concern the values of DAUGHTER HALF-LIFE, which are determined empirically. Finally, as long as it was expected that only fast neutrons would have sufficient energy to produce α emission, there was a value constraint between the value HIGH for the attribute PROJECTILE ENERGY and the value α PARTICLE for the attribute EMITTED PARTICLE (Figure 24).

During the period 1934–1938 much research was directed toward producing transuranic elements, that is, elements with a proton number higher than uranium, the last naturally occurring element in the periodic table. Hence, another important frame shows the attributes and values for DAUGHTER NUCLEUS (Figure 25). One of the attributes of this frame, EMITTED PARTICLE, depends on a physical theory that explains why the only possible values are α PARTICLE, PROTON, β PARTICLE, and NONE. Another attribute, CHEMICAL BEHAVIOR, drawn from chemical theory, takes values that indicate the element used as a carrier when the daughter precipitates from solution. There are strong constraints between the values of the attributes. For example, if an α particle is emitted from a thorium

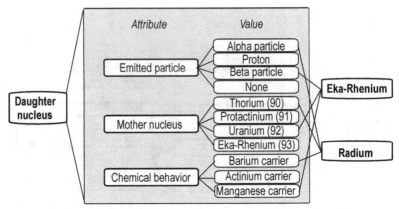

FIGURE 25. Partial frame for 'daughter nucleus'. Two subordinates are shown: eka-rhenium and radium.

mother nucleus, the resulting daughter nucleus will be radium, which has chemical properties similar to barium. The daughter should therefore precipitate in any chemical process that causes the barium carrier to precipitate.

4.3.3. Anomalies in Nuclear Physics during the 1930s

During the period 1934–1938 several groups of scientists worked on induced radioactivity in uranium, among them Fermi's group in Rome, a group in Berlin consisting of Lise Meitner, Otto Hahn, and Fritz Strassmann, and the Curies in Paris. All these groups agreed on how to analyze the different processes that produced transuranic elements by distinguishing contrasting categories from the features we have described and represented in the frame of induced nuclear reactions. We shall now follow just a few of the anomalies that were encountered in the research to exemplify how anomalies and responses to them can be captured in the frame representation. This is not intended as a complete historical account of the research in this period (for detailed accounts see Andersen 1996; Stuewer 1994).

The first anomaly was encountered by the Berlin group (Meitner and Hahn 1936). Primarily from chemical analysis of the decay products, the Berlin team had identified one of the processes as α

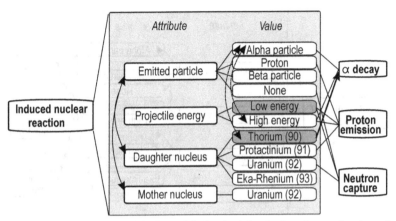

FIGURE 26. Violation of the hierarchical principles during the production of eka-Re by slow neutrons.

emission followed by β decay of the daughter nuclei:

$$^{238}_{92}U + {}^{1}n \rightarrow {}^{235}_{90}Th + \alpha$$

$$^{235}_{90}Th \xrightarrow{\beta} {}^{235}_{91}Pa \xrightarrow{\beta} {}^{235}_{92}U \xrightarrow{\beta} {}^{235}_{93}eka\text{-}Re$$

The end product was a nucleus with the proton number 93. It was expected to be placed in the same group in the periodic table as manganese, technetium (by then called masurium), and rhenium, and it was therefore called 'eka-rhenium' or 'eka-Re' for short. By the same token, it was expected to behave chemically as manganese does and it would therefore be identified in precipitation processes using a manganese carrier.

As explained, it was expected that only fast neutrons could provide sufficient energy to enable the heavy α particle to escape from the nucleus. Hence, in the frame representation there is a value constraint between the value HIGH of the attribute PROJECTILE ENERGY and the value ALPHA PARTICLE of the attribute EMITTED PARTICLE (Figure 26).

However, the process analyzed by the Berlin group could easily be produced when using slow neutrons as projectiles. This meant that, judged from chemical characteristics, the process seemed to be an α emission, but judged from the energy of the projectile, it had to be

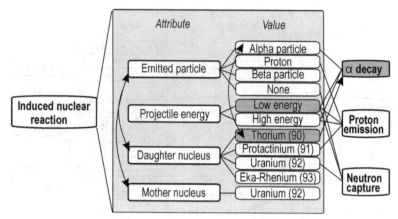

FIGURE 27. Resolving an anomaly by changing a value constraint.

one of the other processes: a violation of the hierarchical principles. In order to resolve the anomaly Meitner and Hahn suggested that the isotope ^{239}U had a very high α instability and that, therefore, neutrons of a much lower energy than were usually necessary to let an α particle escape would cause α decay from this particular isotope. This particular case is therefore not seen as an instance of the normal but as a montrous case for which special rules apply. The anomaly can be interpreted as a monster anomaly that is resolved by giving up one of the characteristics that led to the violation of the hierarchical principles.

In terms of frames, this anomaly was a violation of the hierarchical principles caused by a value of the attribute DAUGHTER NUCLEUS that indicated α DECAY and a value of the attribute PROJECTILE ENERGY that indicated PROTON EMISSION or NEUTRON CAPTURE (Figure 26). This anomaly was resolved by changing the value constraint on the values of PROJECTILE ENERGY for the subordinate concept α DECAY (Figure 27). However, the constraint is not changed in general, but only for this particular isotope, which is expected to have an α instability that is different from the normal. Again, we may interpret the anomaly as a monster anomaly.

Another anomaly encountered by the Berlin team related to a single primary process that they had also identified chemically as producing the daughter nucleus eka-Re. This daughter could only have been produced by a β-emitting uranium isotope. The primary process therefore had to produce uranium, and that requirement excluded both α

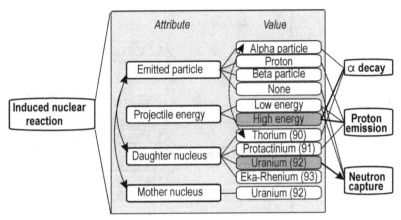

FIGURE 28. Violation of the hierarchical principle during the primary process in eka-Re production.

emission and proton emission. However, it turned out that the process was not intensified when using slow neutrons, and that characteristic also excluded neutron capture (see Meitner and Hahn 1936). Different features therefore excluded different processes in such a way that all known processes were ruled out: in other words, again a violation of the hierarchical principles. Looking at Figure 28 we can summarize the problem by noting that no existing concept allows the following attribute-value pairs: DAUGHTER NUCLEUS: URANIUM and PROJECTILE ENERGY: HIGH. In order to solve this anomaly Meitner and Hahn suggested that the process could be described as an incoming neutron hitting a neutron in the nucleus with enough energy that they both escaped. This suggestion resolved the anomaly by introducing a new category in the contrast set of possible processes. The new category, NEUTRON CHIPPING, did not overlap with any of the three other categories; nor did it fall outside the possible processes defined by Gamow's theory of α emission by quantum tunneling. Hence, the new category could be added unproblematically as a new normal process that might be commonly found.

In terms of frames, this anomaly was a violation of the hierarchical principles since the value of the attribute DAUGHTER NUCLEUS ruled out α DECAY and PROTON EMISSION, while the value of the attribute PROJECTILE ENERGY ruled out NEUTRON CAPTURE (Figure 28). The anomaly was treated as a model anomaly and resolved

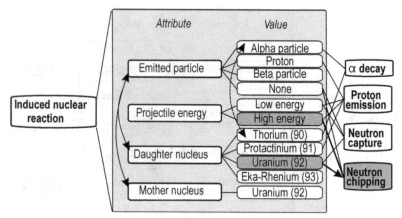

FIGURE 29. Resolving the anomaly by introducing a new subordinate concept, 'neutron chipping'.

by introducing a new subordinate concept, NEUTRON CHIPPING, that had a unique value distribution different from the possible value distributions of the other subordinates (Figure 29). Although, again, a new concept has been introduced and value constraints have been changed in the process of resolving the two anomalies just described, the process has been conservative: no previously classified processes have been reclassified. Although the conceptual structure has changed, the original similarity and difference classes on which it was based have been preserved.

During most of the 1930s, the conceptual structure that was the common property of research groups in nuclear physics, and particularly of the Berlin group, assumed a form that excluded any decay processes except those producing objects the size of an α particle or smaller. For example, one of their close associates, von Weizsäcker, in his book *Die Atomkerne* (1937) discussed all possible induced radioactive processes from a list of all possible combinations of protons, neutrons, deuterons, α particles, and γ radiation as projectiles and decay products. Drawing on our previous discussion, we could therefore say that the process now known as nuclear fission was excluded by the ontological knowledge implicit in frames like Figure 29: no processes except the four listed here were admitted, and these were accepted as exhaustive. The recognition of fission required a change in the conceptual structure of a different sort. Before considering the conclusion of this historical episode, we will review what we have so far established about the nature of conceptual change. We will then consider the topic of nonconservative, or revolutionary, conceptual change by first examining an additional episode from the history of ornithology during the Darwinian revolution, before going on to show how similar mechanisms allow us to understand the difficulties that surrounded the discovery of nuclear fission.

4.4 TYPES OF CONCEPTUAL CHANGE

The case studies in Sections 4.3.1 and 4.3.3 have shown how the hierarchical principles can be violated in different ways. According to the no-overlap principle, it must be possible to assign any object encountered unambiguously to one and only one of the concepts in the relevant contrast set. Hence, the principle is violated if an object is encountered that judged from some features has to be an instance of one concept, but judged from other features has to be an instance of a contrasting concept (Figures 30 and 31).

However, whether an anomaly is a violation of the no-overlap principle or of the exhaustion principle will often be a matter of interpretation. For example, let us return to Figure 26. In this frame uranium as daughter nucleus indicates that the process is neutron capture. But the projectile energy does not have to be low, and that characteristic excludes neutron capture. This anomaly can be interpreted as either a violation of the exhaustion principle since there is no existing concept

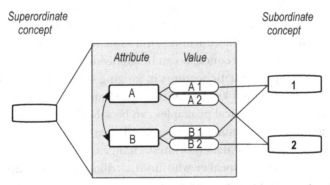

FIGURE 30. A partial frame for a kind hierarchy with two subordinate concepts.

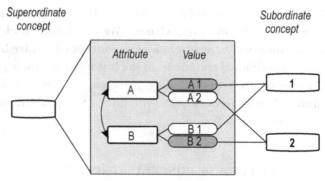

FIGURE 31. Violation of the no-overlap principle in a kind hierarchy with two subordinate concepts. Judged from the value of attribute A displayed by the encountered object, it must be an instance of the subordinate concept 1, but judged from the value of attribute B it must be an instance of the subordinate concept 2.

that captures the combination of a uranium daughter and a high projectile energy or as a violation of the no-overlap principle because one feature (the value of the attribute **DAUGHTER NUCLEUS**) indicates that the process is neutron capture, while another feature (the value of the attribute **PROJECTILE ENERGY**) indicates that the process is α decay or proton emission.

What we can also see from our case studies is that anomalies can be resolved in two ways without changing the overall conceptual structure. For the overall structure to be maintained, the similarity and dissimilarity relations between the concepts in the contrast set must remain unaffected, and only their attachment to the features involved in the violation of the principles may be changed. This can be done in two different ways. First, some of the conflicting features by which similarity and dissimilarity between the concepts in question are judged can be given up. Second, one of the concepts can be subdivided, restricting the relevance of the problematic features in judging similarity and dissimilarity to contrasting subkinds. Hence, in the frame representation violations of the hierarchical principles can be solved by changing value constraints or by introducing new subordinate concepts (Figures 32 and 33). However, the details of these changes may vary between different individuals. A speaker who finds attribute A more important than B in categorization may be inclined to change the value constraints in the way represented in Figure 31. However, a speaker who

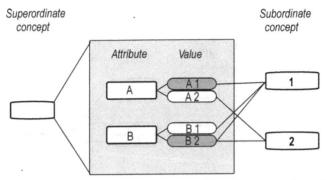

FIGURE 32. Resolution of an anomaly that violates the no-overlap principle by changing value constraints. In this case, a constraint that the value A1 only occurs together with the value B1 has been given up, and instead the value A1 may occur together with either of the values B1 or B2.

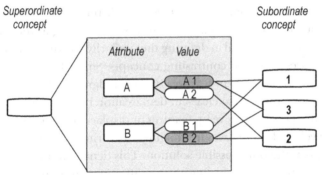

FIGURE 33. Resolution of an anomaly violating the no-overlap principle by adding a new subordinate concept.

finds attribute B more important than A may change other value constraints (Figure 34). In this situation, the first speaker will categorize the object as an instance of concept 1, while the second speaker will categorize the object as an instance of concept 2. Here, differences between the conceptual structures of the two speakers have become apparent through their different categorizations of the same object. However, it is important to note that although they previously categorized all objects in the same way, the difference existed in a latent form through their different emphasis on the attributes.

Neither of the changes represented in Figures 31-34 affects previous results. Objects that have previously been categorized as instances

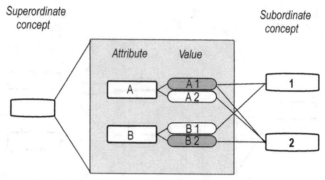

FIGURE 34. Resolution of an anomaly that violates the no-overlap principle by changing value constraints. In this case, a constraint that the value B2 only occurs together with the value A2 has been given up, and instead the value B2 may occur together with either of the values A1 or A2.

of concept 1 or instances of concept 2 will remain classified in the same way. This kind of change is only possible as long as the features involved are not central in defining the similarity and dissimilarity relations between the two contrasting concepts. Since different speakers may emphasize different features in their judgment of similarity and dissimilarity, there is no clear-cut demarcation between those cases in which it is possible to give up features or displace them to the subordinate level, and those cases in which restructuring the two contrasting concepts is the only possible solution. This demarcation may vary for different speakers, according to the difference in graded structures for the concepts in question.

Changing the similarity and dissimilarity relations themselves rather than their attachment to features will change the conceptual structure. Contrary to changing only attachment to features or adding subconcepts, changing the similarity and dissimilarity relations creates incommensurability between the original and the changed conceptual structure. This form of conceptual change will occupy our attention in the last two sections of this chapter.

4.5 REVOLUTIONARY CHANGE

On the basis of the frame model we are now in a position to distinguish among several different types of conceptual change, to explain why some kinds of change are more severe than others, and to explain why those kinds of change have the potential to create communication difficulties. It is important to separate these issues. Although it is still common in the literature to equate incommensurability with communication failure between the supporters of different conceptual structures, this need not always be the case (Hoyningen-Huene 1993: 254–256). As we will see later, incommensurability may occur in situations that do not create communication failure (for example, the replacement of the Sundevall taxonomy), and in situations that create dramatic communication failures (a proposal that might have led to the discovery of nuclear fission well before 1938). We will therefore separate those changes that create incommensurability between conceptual structures from those that create communication difficulties.

We treat incommensurability as a purely conceptual matter. Parallel to the account of the conceptual changes that may occur in normal science, we count any conceptual system as incommensurable with a predecessor if it was created by changing the similarity and difference relations that establish contrast sets. Another way to express this condition is that in the new structure existing entities are redistributed across existing categories, although it is important to acknowledge explicitly that this occurs because the basis for category membership has changed. As we will see, such changes are at best a partial cause of communication failures in science, although, as we will argue later, they are a cause that cannot be dispensed with in understanding the historical development of science. For historical examples of such processes we return to the case studies on ornithology (Section 4.5.1) and on nuclear physics (Section 4.5.2).

4.5.1. The Gadow Taxonomy: Revolutionary Change without Communication Failure

In Section 4.3.1 we explained how the seventeenth-century classification of birds into water birds and land birds dealt with anomalies like the South American screamer. We showed how this anomaly was accommodated by introducing the new category GRALLATORES into the contrast set that previously held only the two categories, WATER BIRDS and LAND BIRDS (Figure 22).

The Darwinian revolution caused radical changes in bird classification. Influenced by Darwin's beliefs that species change over time and

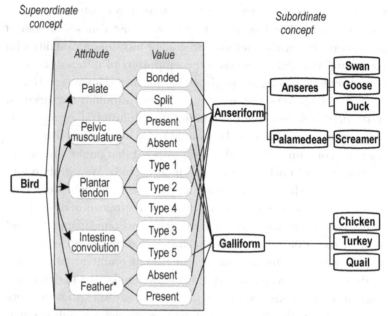

FIGURE 35. A partial frame for Gadow's concept 'bird' and its taxonomy. *The attribute 'feather' refers to the existence of the fifth secondary.

therefore affinity among species must be founded on their common origin, ornithologists realized that many features used as classification standards in pre-Darwinian taxonomies were irrelevant, and they began to search for features that displayed the evolutionary origin of birds. In a popular post-Darwinian taxonomy proposed by Gadow in 1892 (Figure 35), a different set of attributes was adopted (Gadow 1892: 230–256); that included PALATE STRUCTURE, PELVIC MUS-CULATURE FORM, TENDON TYPE, INTESTINAL CONVOLU-TION TYPE, and WING-FEATHER ARRANGEMENT. In the last category, the presence or absence of the fifth secondary feather was of particular interest. Embedded in the Gadow taxonomy is a whole new concept of BIRD. The strong intraconceptual relations among all attributes reflect the assumption that similarities in these anatomical features reveal a common origin, and therefore the values of these attributes ought to be correlated. The strong constraints among the attributes significantly reduce the number of the possible property combinations. For example, the combination PALATE: BONDED and FIFTH SECONDARY FEATHER: PRESENT, exemplified by screamers, becomes impossible, and Sundevall's category GRALLATORES with its subconcept SCREAMER cannot be included in the contrast set at the subordinate level. At the same time, the similarities between water birds and screamers in skull character, skeleton, wing pattern, and feather structure suggested that they should be put under the same covering concept. Consequently, Gadow introduced a new subordinate concept, ANSERIFORM, to denote both waterfowl and screamers.

The frame representation shows why the pre- and the post-Darwinian taxonomies were incommensurable and confirms Kuhn's account of how incommensurability arises. As a result of addition, deletion, and rearrangement of kind terms, a holistic redistribution of referents occurred. Because of the referent redistribution, many terms in the new taxonomy could not be translated to the old ones, nor the other way around. Consequently, it became possible but not inevitable that communication between followers of the two systems would be impeded. For example, the followers of the Sundevall taxonomy might regard Gadow's category ANSERIFORM as confusing because they could not find an equivalent term without violating the no-overlap principle. Within the Sundevall taxonomy, the no-overlap principle requires that no grallatores are also natatores. The referents of Gadow's ANSERIFORM overlap those of Sundevall's NATATORES – the former includes the latter as a subset, but they are not in species-genus relation. The followers of the Gadow taxonomy, on the other hand, might regard Sundevall's GRALLATORES as incomprehensible because of its overlap with ANSERIFORM.

Despite these differences, the historical confrontation between the Sundevall and Gadow taxonomies was not marked by failure of communication. Although, at first glance, the attribute list embedded in the post-Darwinian Gadow taxonomy is considerably different from the one in the pre-Darwinian Sundevall taxonomy, notice that these two lists of attributes are compatible: none of the attributes listed in one taxonomy introduces a violation of the no-overlap principle for attributes already used in the other. A closer examination of these attributes further shows that the two lists of attributes are similar – all of them are anatomical parts of birds.

Historical evidence indicates that the two rival taxonomies were compared and evaluated despite their incommensurability. Although there were debates regarding the merits of the two rival systems, criticisms from either side were mainly based upon observations of similarity and dissimilarity relations between birds. The main objection to the pre-Darwinian taxonomy was, for example, that it grouped many dissimilar birds together (Newton 1893). Presented with compelling evidence in the form of generally accepted similarity and dissimilarity relations, the community quickly formed a consensus. Before the end of the nineteenth century, the Gadow taxonomy was accepted by the ornithological community (Sibley and Ahlquist 1990).

The replacement of the Sundevall taxonomy by Gadow's alternative is an example of revolutionary change in which there is no major failure of communication, and in which rational comparison of the incommensurable positions occurred. The pre- and post-Darwinian taxonomies specify different similarity relations. The former put SCREAMER and the equivalent of WATER BIRD under two contrastive covering terms and emphasized their dissimilarity, while the latter put them under the same covering term and emphasized their similarity. However, the different but compatible lists of attributes embedded in the pre-Darwinian and post-Darwinian taxonomies provided a basis for communication between the two positions and, in the end, a platform for rational comparison. Because the attribute lists were compatible, people from both sides could agree with each other on what attributes should be counted as relevant in judgments of similarity. When observations showed more and more similarities between screamers and waterfowl in skull character, skeleton, wing pattern, muscular system, and digestive system, supporters of the pre-Darwinian taxonomy had to agree that all these similarities were relevant and accept them as legitimate evidence for testing their taxonomy. When observations of the similarities between screamers and water birds became overwhelming, they had no choice but to admit that their taxonomy was defective and adopt Gadow's alternative. This case shows clearly the features that Kuhn insisted upon in his own mature account of incommensurability: failure of communication during revolutionary change is at best partial and may not be present at all, while incommensurability need not preclude rational comparison of rival positions. But there are historical cases in which incommensurability and communication failure are linked. The delayed discovery of nuclear fission is an example.

4.5.2. Noddack, Fermi, and Fission: Revolutionary Change with Communication Failure

In Section 4.3.2 we described the kind hierarchy of induced disintegration processes that were the key categories in the research on transuranic elements in the mid-1930s. In Section 4.3.3 we also described some of the anomalies that appeared in this research and showed how these were accommodated by introducing new categories into the contrast set of possible disintegration processes or by changing value constraints for existing concepts.

However, a more radical change was suggested by the German chemist Ida Noddack. From the outset she questioned the identification of transuranic elements and suggested that they could instead be fractions of a nucleus that had exploded (Noddack 1934b). As part of her argument, Noddack first questioned whether the new transuranic element number 93 would have the chemical characteristics that Fermi and his group had assumed when they did the chemical investigations. To analyze Noddack's objection we have to look at part of the frame for CHEMICAL ELEMENT (Figure 36) that underlies the individual values of CHEMICAL BEHAVIOR in the frame for daughter elements (Figure 25). The daughter nuclei were identified chemically by precipitation processes. In these processes the different possible daughter elements would precipitate with elements from the same subgroup, represented by a column of elements in the periodic table, shown as values of the attribute PRECIPITATION PROCESS in Figure 36. Following the periodic table as it appeared in 1934 (Figure 37), for example, the element radium would precipitate with barium (column 2). Likewise, since Fermi and his collaborators expected that element 93 would be placed just after element 92, element 93 should precipitate with rhenium and with manganese (column 7). Noddack argued that at least fourteen other elements would also precipitate with manganese. Further, she was not sure whether element 93 would actually have chemical properties that would make it precipitate with manganese and rhenium. Hence, for Noddack the chemical properties did

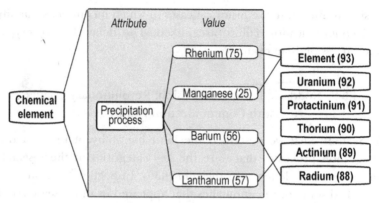

FIGURE 36. Partial frame for 'chemical element'.

							Periodsches System der Elemente										
1	2	3	4	5	6	7	8	9	10	11	12	13	14	15	16	17	18
																1 H	2 He
3 Li	4 Be											5 B	6 C	7 N	8 O	9 F	10 Ne
11 Na	12 Mg											13 Al	14 Si	15 P	16 S	17 Cl	18 Ar
19 K	20 Ca	21 Se	22 Ti	23 V	24 Cr	25 Mn	26 Fe	27 Co	28 Ni	29 Cu	30 Zn	31 Ga	32 Ge	33 As	34 Se	35 Br	36 Kr
37 Rb	38 Sr	39 Y	40 Zr	41 Nb	42 Mo	43 Ma	44 Bu	45 Rh	46 Fd	47 Ag	48 Cd	49 In	50 Su	51 Sb	52 Te	53 J	54 X
55 Cs	56 Ba	57 * La	72 Hl	73 Ta	74 W	75 Re	76 Cs	77 Ir	78 Pi	79 Au	80 Hg	81 Tl	82 Pb	83 Bi	84 Pe	85 --	86 Rd
87 --	88 Ra	89 Ac	90 Th	91 Pa	92 U	93 --	94 --	95 --	96 --								

		58 Ce	59 Pr	60 Nd	61 --	62 Sm	63 Eu	64 Gd	65 Tb	66 Dy	67 Hu	68 Er	69 Tu	70 Yb	71 Cp

FIGURE 37. The periodic table of the elements in 1934. Comparison with modern symbols: Re = rhenium, Ac = actinium.

not point to element 93, but rather to many other elements, among them several light elements (Figure 38).

Noddack's chemical argument falls nicely within the range of the frame for CHEMICAL ELEMENT. However, it has implications for the interpretation of the nuclear processes that cannot be accommodated within the frames for DAUGHTER NUCLEI and INDUCED NUCLEAR REACTION. As described in Section 4.3.2 the frame for DAUGHTER NUCLEI has attributes derived from chemical theory as well as attributes derived from physical theory. The latter attributes

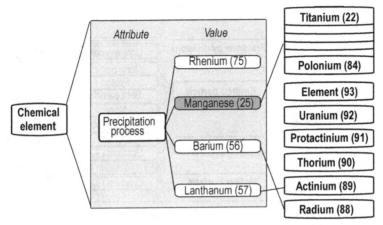

FIGURE 38. For the chemist Ida Noddack the precipitation with manganese did not point to element 93, but rather to a long list of light elements.

can only accommodate daughter nuclei that occur close to the mother nucleus in the periodic table (Figure 25). In this frame there is no room for a light daughter of a heavy mother nucleus.

Similarly, as described in Section 4.3.2 the frame for INDUCED NUCLEAR REACTION has a structural invariant linking three attributes: EMITTED PARTICLE, MOTHER NUCLEUS, and DAUGHTER NUCLEUS. This structural invariant reflects Gamow's theory of decay, which taught that only particles up to the size of the α particle can escape the nucleus (as described in Section 2.8). Hence, as a result of the structural invariant among EMITTED PARTICLE, MOTHER NUCLEUS, and DAUGHTER NUCLEUS (Figure 26) a daughter nucleus produced from a heavy element simply cannot be a light element.

To account for the possible production of light elements Noddack suggested that they could have been produced by the division of the nucleus into several large fractions. However, this was a different way of conceiving of the fate of the nucleus. According to physical theory, nuclear disintegrations had to happen by either α emission, proton emission, neutron capture, or β emission. In all cases one heavy nucleus would transmute into another heavy nucleus by releasing a small particle. This is represented in Figure 26 through the attributes MOTHER NUCLEUS, DAUGHTER NUCLEUS, and EMITTED PARTICLE, which all have connected and restricted ranges of

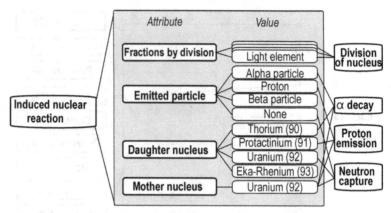

FIGURE 39. Modified frame for 'induced nuclear reaction' including the new subordinate 'division of nucleus', required to accommodate Noddack's suggestion.

values. In this frame there is no room for a process in which a heavy nucleus divides into several light nuclei. To accommodate Noddack's suggestion a whole new attribute would have to be included, FRACTIONS BY DIVISION (Figure 39), but such an attribute seemed to be precluded by the existing conceptual structure of physical theory.

Nobody in the scientific community ever reacted publicly to Noddack's suggestion. Apparently, her suggestion could not be taken seriously. To explain this difference between Noddack and other scientists, we must look at the severity of the anomaly that led to the suggestion. In our terms it would be necessary for Noddack to present an anomaly so severe that it would motivate substantial changes of the conceptual structure. But unfortunately no one else could see the anomaly.

Noddack was an analytical chemist. She had worked for years on the gaps in the periodic table. Earlier in 1934 she had expressed her firm belief that transuranic elements probably existed, but that accurate predictions of the characteristics of the transuranic elements had to be made before they could be discovered (Noddack 1934a: 304). In the same paper she described constraints on chemistry derived from theoretical physics as 'dogmas' that would one day be refuted. For her, chemical identifications clearly carried more weight in identifying elements than physical expectations of possible decay series. By the same token, if chemical characteristics suggested that a new disintegration process had to be added, then so be it. For Noddáck, the cost of solving a chemical problem would be giving up a mere presupposition about what might or might not exist in an area of research that she had not entered before.

Fermi's team, on the other hand, used the conceptual scheme of disintegration processes to narrow the range of possible elements that might have been created in their experiments and then made chemical analyses only within this narrow range of possibilities. To them, as well as others in the field, Noddack had not pointed to any serious anomalies, but only to "a lack of rigor in the argument" (Amaldi 1984: 277). This was definitely not enough to trigger a fundamental change in the conceptual structure.

However, four years later another anomaly did lead to fundamental change. The Berlin group was examining a process in which radium was produced by two sequential α decays, before undergoing sequential β decays yielding actinium and thorium (Hahn and Strassmann 1938):

$$^{238}_{92}U + {}^{1}n \rightarrow {}^{235}_{90}Th + \alpha$$
$$^{235}_{90}Th \longrightarrow {}^{231}_{88}Ra + \alpha$$
$$^{231}_{88}Ra \xrightarrow{\beta} {}^{231}_{89}Ac \xrightarrow{\beta} {}^{231}_{90}Th$$

In the analysis of the process radium had been identified through precipitation with barium as the carrier element. However, in December 1938 Hahn and Strassmann discovered that they could not separate the radium from its barium carrier. In the frame representation we see that Hahn and Strassmann examined another attribute in the partial frame, the behavior in a further chemical separation, and this additional attribute revealed a violation of the hierarchical principles. From a nuclear physics viewpoint the element had to be a heavy element close to uranium in the periodic table, but from a chemical viewpoint it seemed to be the light element barium (Figure 40). However, there was no way that barium could be added as a subconcept to the frame of daughter nuclei. Barium could not be produced from a heavy nucleus by emission of single α particles or protons. Attributes completely different from emission of small particles are needed to account for the production of light elements. Thus, in order to allow for barium as a decay product, it is necessary to restructure the frame

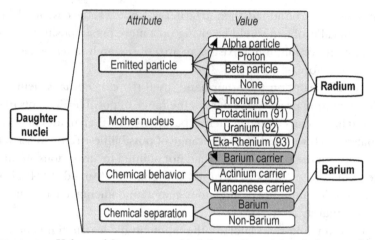

FIGURE 40. Hahn and Strassmann added the attribute 'chemical separation' to the frame for 'daughter nuclei', revealing a violation of the hierarchical principles.

of daughter elements into a frame for decay products with two main kinds: transmuted nuclei similar to the previous daughters and divided nuclei that can be light fractions (Figure 41). Gamow's theory of decay had explained the range of values of the attribute EMITTED PARTI-CLE, but scientists soon realized that a theory of the nucleus previously advanced by Bohr (Bohr 1936) could explain how a division into several large parts might occur (Meitner and Frisch 1939; Bohr and Wheeler 1939). This new conceptual hierarchy had far-reaching consequences for all the previous results on transuranic elements. The new hierarchy also included the production of transuranic elements, but the value distribution of these categories was no longer a settled question, and if they changed, the previous categorizations could no longer be maintained. Within a few months the Berlin group recategorized their previous results as fission and retracted their results on transuranic elements (Meitner and Frisch 1939b; Hahn and Strassmann 1939). Similarly, Fermi added a footnote to the Nobel Prize Lecture he delivered after receiving the prize for his work on induced radioactivity, that the new discovery made it necessary to reexamine all previous results on transuranic elements (Fermi 1939).

The new conceptual hierarchy gave rise to new puzzle-solving activities. The splitting of a heavy nucleus into two light nuclei led to a

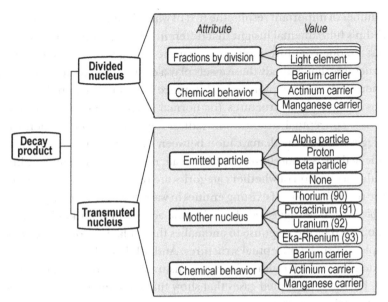

FIGURE 41. Partial frame for 'decay product'.

considerable excess of neutrons in these two nuclei that had to be accounted for. The liberation of neutrons thus became one such new puzzle, raising new research questions about the number of neutrons liberated and the processes that liberated them (von Halban, Joliot, and Kowarski 1939a, 1939b). Other previously established results were lost, namely, the results on transuranic elements (Feather and Bretcher 1939). New categorizations of all the previously examined processes had to be made, now using measurements of how decay fragments recoil for identifying transuranic elements (McMillan 1939), a feature that had not been considered before at all.

4.6 CONCLUSION: A PLACE FOR THE COGNITIVE HISTORY OF SCIENCE

Throughout the course of this chapter we have recovered central features of Kuhn's account of scientific change in terms of the theory of concepts that he developed in his mature work and parallel work in cognitive science that has provided a means for describing the details of conceptual structures as Kuhn conceived them. We have established a number of important results, the first of which is simply a vindication of Kuhn's fundamental insight that different scientific communities possess unique cognitive assets, and that the changes in the conceptual structures that embody these assets play a central role in the history of science. We have explored the difference between research that conserves conceptual structures (or normal science) and research that revises conceptual structures in radical ways (revolutionary science), suggesting that the demarcation between these two patterns of scientific activity is a question of neither scale nor historical frequency but rather a question of whether categories are revised in ways that require the reclassification of existing entities in ways that were formerly impossible. We have suggested that scientific activity of both kinds can be motivated by the response to anomalies that violate the ordering principles for such conceptual structures. And, while we have deferred an extended consideration of incommensurability for the next chapter, we have also presented cases that show that incommensurability does not entail communication failure between supporters of alternative positions. But our main conclusion links the evidence presented in the account of anomalies in normal science during the 1930s with the revolutionary change brought about during the discovery of nuclear fission in the winter of 1938. The conclusion we wish to urge is that cognitive factors are ineliminable in reaching a historical understanding of this case, whatever use is made of other historical causes.

Ida Noddack did not discover nuclear fission. Rather she made a suggestion in 1934 that today would be understood as requiring a fission process but could not be so understood at the time. It is this cognitive aspect of the situation, the impediments to understanding Noddack's suggestion as entailing fission of the nucleus, that frame analysis illuminates. As a matter of record, Noddack was not credited with discovering fission either before or after the events that today are counted as the discovery.

Noddack's background and disciplinary affiliation were in chemistry. Her research was directed to filling in blanks in the periodic table, an activity that gave paramount importance to the determination of the precise chemical properties of particular elements. This disciplinary background and research focus go some way to explaining both her skepticism about the Fermi group's claim to have produced element 93 and her readiness to consider possibilities that the Fermi group disallowed.

Fermi's research group; Hahn's group in Berlin (including Lise Meitner), the Paris group, including Joliot and Curie, and others elsewhere were composed of people with backgrounds in both physics and chemistry. In contrast to Noddack, their research focus was induced nuclear disintegration. These groups followed each other's work closely and according to our analysis shared a common conceptual structure for understanding decay processes that was not shared with Noddack. The research interests of people like Noddack and the members of the Fermi-Hahn consortium overlapped in the area of transuranic elements, but the two groups had different conceptual resources.

It is not sufficient to explain the response – or rather lack of response – to Noddack's proposal of 1934 to point out the difference between these research groups, even if we add that Noddack was a woman in a field dominated by men. As a matter of fact this particular field was home to several prominent women, for example, Curie and Meitner. Clearly their careers were affected to varying degrees by male perceptions of the status of women scientists. Noddack also had considerable credibility within the wider scientific community: she had been awarded the Justus Leibig Medal by the German Chemical Society in 1931 and the Scheele Medal from the Swedish Chemical Society in 1934. She was a leading member of the team that discovered rhenium (element 75), work that was nominated (unsuccessfully) for a Nobel Prize several times during the 1930s. On the negative side, her parallel claim to have detected element 43 became controversial.

Although male critics might well have left unstated any objections to her proposal on the grounds that she was female, they might possibly have objected that she was a chemist making proposals about a subject outside her area, or a nonspecialist venturing on specialist territory. Actually, Hahn also considered himself a chemist (see the following excerpt). With hindsight, what is striking is that there are no such rebuttals – there is simply *no* public response at all. Neither Fermi, nor Hahn, nor Joliot and Curie (to mention only the most obvious candidates) felt obliged to say anything public about Noddack's proposal.

However, in retrospect, appraisals of Noddack say a number of interestingly similar things. Writing to Strassmann in 1939 about an article by Noddack that appeared in *Naturwissenschaften,* Hahn says:

In 1934, because of the theoretical conceptions and experimental results, the world's leading nuclear physicists of the time were not capable of predicting the present results [*heutigen Ergebnisse,* i.e., the Meitner explanation of the barium anomaly]. There was for us as chemists no reason to doubt the claims of physics. (Krafft 1981: 319; a photograph of the relevant portion of the letter appears on p. 320)

And in 1946 Hahn wrote:

From another side (Ida Noddack) the objection was made that one had even to rule out all elements in the periodic table before one could make the claim to have an element 93. At that time this objection was never seriously discussed, as contradicting all physical conceptions about nuclear physics. (Hahn 1946: 253)

A historical summary from Treumann is particularly revealing:

Moreover there were two or three suggestions in the literature that fission might be possible in neutron-heavy nucleus interactions, one by Ida Noddack, the other by v. Grosse. But these suggestions had, before the discovery of fission, been ignored by the whole scientific community; . . . After fission was discovered and interpreted as such by Meitner and Frisch, after the full theory of fission had been developed by Bohr and Wheeler in the approximation possible then, these suggestions were ignored as well. The reasons for this have never been illuminated and may be difficult to reconstruct, but one of them may be found in the lack of a theoretical model underlying the suggestion. (1991: 144ff.)

The quotations from Hahn document the reluctance to acknowledge Noddack's proposal as a prediction of fission before or after the work of Hahn, Strassmann, Meitner, and Frisch in the winter of 1938. And they also point to a cognitive, rather than a purely social, basis for this lacuna. Although we would expect scientists to try to present cognitive grounds for ignoring Noddack in later writings, even if the major historical factors operative at the time deserve to be classified as social, we should still ask whether these later cognitive criticisms have a real historical basis. The scientists are handicapped by their vocabulary of appraisal, which extends little beyond absence of suitable 'experimental results' or 'lack of a theoretical model'. We are in a position to give a much more detailed explanation.

Until the winter of 1938 induced nuclear reactions were understood by means of a conceptual structure that we have represented by means of the frame diagrams in Figures 27 and 28. This initially allowed only three subconcepts: α decay, proton emission, and neutron capture, later augmented with neutron chipping. This limited range of subconcepts came about primarily because of the strong constraint in the frame among the attributes MOTHER NUCLEUS, DAUGHTER NUCLEUS, and EMITTED PARTICLE. The attribute constraint reflected a specific subsidiary theory: Gamow's theory of α emission by tunneling through the potential barrier around the nucleus. The addition of the new subconcept NEUTRON CHIPPING had not altered this fundamental feature of the frame; indeed, the attribute constraint was active during the introduction of the new subconcept. From the viewpoint of this conceptual structure, Noddack's proposal attempted to introduce a new subconcept that violated this attribute constraint, but without giving any grounds for abandoning the constraint. As we have seen previously, such grounds would be, for example, an anomaly that violated the hierarchical principles we have described in Section 4.2. So Noddack's proposal was cognitively defective on two scores: it only made sense as a dramatic revision of the existing conceptual structure, and it provided no appropriate motivation for making such a revision. The response of researchers using the conceptual structure we have described was that the proposal made no sense; put in more traditional terms: what Noddack was proposing was unthinkable to anyone using the old conceptual structure.

Only when Hahn and Strassmann provided a reason for attacking the attribute constraint that had previously existed in the frame for induced nuclear reactions could a new structure emerge in which fission was possible. And the new emission possibility represented by fission had to be legitimated by introducing Bohr's liquid drop model of the nucleus, in the same way that Gamow's tunneling theory of α emission had supported the emission possibilities in the older conceptual structure. Bohr's liquid drop model now supplied the 'lack of a theoretical model' of which Treumann later complained; however, it is important to see that the theory legitimates the new conceptual structure, rather than just providing an interpretation for Hahn's experimental results. But this change in conceptual structure had not been brought about by Noddack, and consequently she received little retrospective credit, even though she could now be understood, retrospectively, as proposing what came to be known as fission of the nucleus *after* the new conceptual structure had been generally adopted.

If Fermi, Hahn, and similar researchers were impeded from understanding Noddack's suggestion by the factors we have analyzed, why was Noddack able to make it? She must have been aware of the general pattern of the conceptual structure for induced nuclear reactions, that is, the list of allowed products in artificially induced nuclear disintegration. But she seems to have been either unaware of the attribute constraint that generated the specific list of subconcepts accepted before winter 1938 or less convinced of the theoretical explanations that suggested why the constraint had to hold. The constraint was based on Gamow's theory of α emission.

We do not discount the role of other historical factors, including social factors, in the reception of Noddack's proposal, although neither her sex nor her disciplinary affiliation seems strongly implicated. The most likely social factors predisposing members of the Fermi group, at least, against suggestions from Noddack lie in the competition between Noddack and Italian researchers to claim the identification of element 43. Her claim to have detected the element, which she named 'masurium', as early as 1927, was ultimately rejected in favor of work done in Italy by Carlo Perrier and Emilio Segré in 1937. They named the element they had isolated 'technetium'.

Even if social factors and other historical causes contribute to the explanation of the fate of Noddack's proposal, the cognitive factors we have indicated are ineliminable. Whatever other motivations researchers had for suspecting or rejecting suggestions made by Noddack, the point that they simply could not understand the proposal using their current conceptual structure presents an already insurmountable objection to further consideration. The key feature revealed by our analysis is the constraint relation among the attributes in the frame for induced nuclear reactions. That the researchers in question operated within a conceptual structure with these features is ` shown both by the responses to anomalies and by the way in which the frame was finally revised to accommodate fission.

Our analysis concludes that the response to Noddack was not primarily social resistance, but rather cognitive incomprehension. The analysis explains both the curious silence about Noddack's proposal before 1938 and the failure to give her retrospective credit after the event. In Kuhn's original terms we could say that Noddack was proposing a revolutionary change in the paradigm without providing an anomaly competent to create a crisis state in which a new alternative could mature. In terms of our cognitive account she can be seen to be making a suggestion that is nonsensical to people using the conceptual structure that contains the attribute constraint we have described. The modification of that constraint came about because of an anomaly (the Hahn-Strassmann barium result) that Noddack had no hand in creating and led to a revision of the conceptual structure by Meitner and Frisch (through the deployment of Bohr's liquid drop model) for which Noddack could claim no credit. Before and after 1938 researchers had cognitive grounds for their silence about Noddack, which we have now been able to specify in detail.

Like all major revolutions, the discovery of fission led to the disappearance of entities previously accepted as existing, in this case, the transuranic elements. Although Fermi received a Nobel Prize for his work in this area, after Meitner and Frisch introduced Bohr's liquid drop model to explain Hahn and Strassmann's barium anomaly, the claims to have produced transuranics before 1939 were dropped. The first production of transuranic elements was credited to other workers elsewhere, for example, Seaborg in California in 1941. The processes by which the new elements were created were understood in terms of the conceptual structure that existed after the changes brought about by Meitner and Frisch. Changes like these are evidence of the creation and elimination of opportunities to categorize entities that we have already suggested as the main characteristic of revolutionary change. Such changes also create the phenomenon of incommensurability between the old and new conceptual systems, and it is to a consideration of this issue that we now turn.

5

Incommensurability

5.1 INTRODUCTION

In this chapter we will use the methods introduced in previous chapters to clarify and extend Kuhn's mature account of incommensurability. We will show that incommensurability between conceptual structures is created by changes that are neither total nor instantaneous. We will also draw out various conclusions that Kuhn suggested but did not elaborate, for example, that incommensurability varies in degree or importance and that the degree correlates with the position of a concept in a hierarchy or conceptual structure as depicted by the corresponding frame. Throughout the next two chapters our main historical focus will be the Copernican revolution, an episode that Kuhn never treated satisfactorily (Westman 1994; Barker 2001). On the basis of our new account, we will suggest that incommensurability may occur even within what Kuhn and earlier writers have regarded as a single paradigm and that this kind of conceptual difficulty may in itself be a motive for conceptual revision. As we will see in Chapter 6, one of the most important motives for Copernicus' revision of Ptolemaic astronomy was a problem of just this type.

5.2 THE DEVELOPMENT OF KUHN'S CONCEPT OF INCOMMENSURABILITY

Kuhn significantly refined his philosophical account of science in the years after the publication of *The Structure of Scientific Revolutions*. The concept of incommensurability also underwent major revisions. From an initial description that emphasized similarities to visual gestalt switches, Kuhn moved in the 1980s to an account that described incommensurability in solely linguistic terms. In the 1990s he further refined this account by limiting the nature of the terms and conceptual structures in which incommensurability appeared. These changes attempted to limit an account that had been misread as global to one that was clearly local. However, on the basis of the cognitive rereading of Kuhn's concept of incommensurability, we will suggest that although incommensurability is created locally and has local effects, it is the result of the operation of mechanisms that are universal. According to the account of human concepts developed by Kuhn, and in parallel by cognitive psychologists, incommensurability is always a possibility in the development of any human conceptual structure.

The presentation of incommensurability in *Structure of Scientific Revolutions* was strongly influenced by Kuhn's acquaintance with gestalt psychology, although there is also a connection with Wittgenstein. In the *Philosophical Investigations*, a new book when Kuhn was writing *Structure of Scientific Revolutions*, Wittgenstein had used duck-rabbit figures in his discussion of "seeing as" (*Philosophical Investigations* XI: 193–229, esp. 194). Kuhn already had an established interest in psychology when he encountered Wittgenstein's work. In *Structure of Scientific Revolutions* he took over Wittgenstein's examples and used the psychological concept of a gestalt switch to try to explicate the changes that occur when scientists abandon one conceptual structure in favor of another (Kuhn 1970a: 62–64, 122).

Ironically, Kuhn's success in explaining his new concept led to misunderstandings that persisted for decades. The idea of a gestalt switch and the illustrations in terms of duck-rabbit figures were dramatic and easy to understand, but misleading in crucial respects. Kuhn's readers seized on two key aspects. First, during a gestalt switch the entire visual field is reconfigured in a way that excludes the previous configuration from cognition. Second, this change occurs instantaneously. They concluded that the concept Kuhn was explicating – incommensurability – must be marked by similar global changes in conceptual structures before and after a scientific revolution and that these changes must happen instantaneously. These implications contributed to the myth that there was total incommensurability between successive paradigms and total communication failure between their supporters. An additional difficulty was that gestalt switches happen in the mind of individuals, obscuring Kuhn's clear message that the community, not the individual, is the bearer of scientific knowledge, and the locus for change during scientific revolutions.

As soon as these misreadings became apparent, Kuhn denied that his concept of incommensurability was total or that he had claimed total communication failure between supporters of successive paradigms (Kuhn 1974, 1991; Hoyningen-Huene 1993: 206–222). To preclude further misunderstandings he dropped references to gestalt switches and the visual consequences of scientific revolutions. In their place he developed the account of the relations between incommensurable concepts begun in *The Structure of Scientific Revolutions*. He now suggested that the communities of scientists supporting rival paradigms are like different linguistic communities (Kuhn 1970a: 198; Hoyningen-Huene 1993: 212 ff.). The question of the extent and nature of incommensurability could then be addressed by analogy with questions of the extent and nature of translation between natural languages. Incommensurability now became a failure of translation, which naturally limited its scope. Taking real human languages as a model, it was no longer plausible to suggest that a failure of translation at one point or a failure connected with a single activity entailed complete untranslatability of one language into another. It became plausible to confine the source of untranslatability to a particular problematic topic or activity while acknowledging that it might be possible to produce perfectly adequate translations between the same pair of languages in connection with many other activities. In this way Kuhn made plausible his suggestion that although successive paradigms might be incommensurable in some aspects, enough common features would remain to allow a basis for communication between the communities supporting them and possibly furnish a basis for some form of appraisal. However, eliminating the gestalt analogy eliminated a clear – albeit misleading – explanation for the origin of incommensurability. An additional problem was that partially untranslatable human languages developed by parallel historical processes, at the same time, but in communities that were isolated from one another. On the other hand, new paradigms with partially untranslatable conceptual structures developed not at the same time, but sequentially, and in communities that were in close contact or initially identical.

In the last decade of his life Kuhn refined his account by specifying a mechanism that would generate incommensurability within an individual language. The terms that generate incommensurability, he now claimed, were only a subset of the vocabulary of science, specifically terms designating 'natural kinds' like 'gold' or 'poison' (Kuhn 1991: 4). These terms did not appear as a flat database of categories but formed natural hierarchies. The lowest level in a kind hierarchy consisted of concepts constituted by similarity and difference relations learned by extension, according to the theory of concepts Kuhn consistently developed and adhered to throughout his career (see Chapter 2). Changes in these similarity and difference relations would count as changes in the objects at the lowest level of the kind hierarchy. When such a change appeared – as a response to an anomaly perhaps – it might require the revision of kind terms at higher levels in the hierarchy. What had been a single conceptual structure now existed in two versions: the hierarchy before modifications of its lowest level and the hierarchy with the modifications of similarity and difference classes and corresponding changes in the objects that could be accommodated by the natural kinds it tabulated. But a kind hierarchy is a tree structure. The changes introduced by revision in the similarity and dissimilarity relations might be confined to the end of one branch, without causing revisions to high-level concepts in the hierarchy. This account presented the dual aspects of local incommensurability: partial or total failure of translation might occur between communities trying to talk about the subject matter represented by the altered branch while communication continued without difficulty on any topic requiring the use of vocabulary from the unchanged portions of the hierarchy.

Although Kuhn restricted his discussion to scientific categories, it should be clear that all human languages can be reconstructed as incorporating kind hierarchies. So any cognitive problem brought about by the revision of such hierarchies is likely to appear not merely in science but quite generally. As we will see in the next section, the account of concepts developed by cognitive psychologists strongly supports Kuhn's views on these matters, and the particular model that we have adopted, the dynamic frame, can be used to explain his concept of incommensurability in greater detail, and with greater generality, than Kuhn's final discussion in terms of kind hierarchies.

5.3 REPRESENTING INCOMMENSURABILITY IN FRAMES

In his mature work Kuhn redrew the picture of scientific revolutions (Kuhn 1983a, 1991). Changes in taxonomy now captured the revolutionary features of paradigm shifts, and the most important changes during scientific revolutions were conceptualized as taxonomic shifts. However, not all changes in taxonomy are revolutionary. As explained previously (Sections 4.2ff.), revolutionary changes always introduce violations of the hierarchical principles for the categories of the previous taxonomy. This will also be true in cases of the most interesting sort of mismatch between taxonomies, called incommensurability.

As noted earlier (Section 3.4) from a cognitive point of view, a taxonomy is a specific structure in the conceptual field defined by a frame. Generally speaking, then, the changes brought about by revolutions, including incommensurability, may be represented as discrepancies between frames drawn before and after the revolution.

Consider the category of 'physical object'. Before Copernicus, this category divided into mutually exclusive terrestrial and celestial subclasses with opposite features. There are many different ways in which this division might be represented, but, as an example of our general technique, let us consider just two attributes of physical objects: whether they can change and what are their natural motions. In 1500 European natural philosophers generally agreed that celestial bodies were unchanging and moved naturally in circles, while terrestrial bodies were changeable and moved naturally in straight lines (Figures 42 and 43). Note in particular the double-headed arrows linking values in Figure 43. These mean that any object deemed unchanging *must* move naturally in a circle, and any object that moves in a straight line *must* be deemed capable of change.

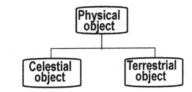

FIGURE 42. Taxonomy for 'physical object', circa 1500.

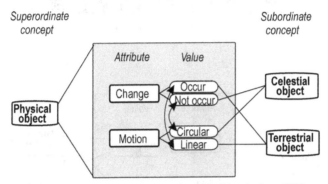

FIGURE 43. Partial frame for 'physical object', circa 1500.

It is not entirely clear how to construct a taxonomy or frame for the equivalent concept after the publication of Newton's work. But at least it is clear that no such bifurcation of celestial and terrestrial objects occurs in the category of 'physical object' after Newton (Figures 44 and 45). One approach would be to say that a physical object is one that obeys the law of universal gravitation, or Newton's laws of motion, or both. What Newton called 'gross matter' obeys both, but light and the various ethers that Newton considered over the course of his career may obey the laws of motion but not the law of universal gravitation. Any of these three forms of matter may be found in the heavens or on earth. The division between celestial and terrestrial objects depends upon where one draws the boundary: perhaps the top of the atmosphere or the orbit of the moon. The category CELESTIAL OBJECT in 1500 corresponds, more or less, to the category of object studied in astronomy after 1700, composed of gross matter and moving about centers of gravitational attraction other than the earth (the moon being an obvious exception).

If we use this representation, the kind hierarchies for 'physical object' before Copernicus and after Newton are not isomorphic, and

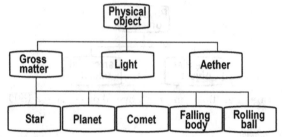

FIGURE 44. Taxonomy for 'physical object', circa 1700.

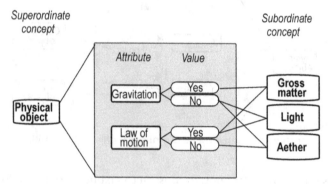

FIGURE 45. Partial frame for 'physical object', circa 1700.

neither are the corresponding frames. However, there is a further important difference. When Galileo performed his experiment rolling balls down inclined planes, he concluded that a ball rolling off an incline onto the surface of the earth would continue to move indefinitely along a great circle (ignoring friction). This is a terrestrial object that continues to move in a circle, a combination forbidden in the frame for 1500. Similarly, Kepler considered comets to be celestial objects that moved in straight lines. These would also be forbidden in the frame for 1500, but allowed in 1700. So not only are the frames for PHYSICAL OBJECT in 1500 and 1700 different in structure, but the differences permit violations of the no-overlap principle applied to the earlier frame. Hence the concepts of PHYSICAL OBJECT represented by the two frames are incommensurable.

As the next step toward a more realistic historical treatment of the Copernican revolution, let us select for special attention the conceptual field defined by the frame of CELESTIAL OBJECT according to seventeenth-century astronomy (Figure 46). This frame represents

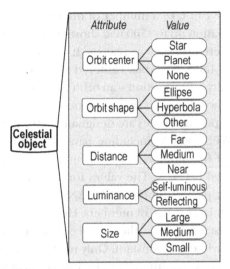

FIGURE 46. Partial frame for 'celestial object', circa 1700.

a conceptual structure widely shared by mechanical natural philosophers in the early eighteenth century but by no means universally
accepted. While supporters of the "new philosophy" would find the
concepts laid out in this frame familiar and acceptable, the wider
community of scholarship still included many people who accepted
Aristotelian or Tychonic views of the world that would be incompatible with much or all of this structure.

On the left we see a single node designating the superordinate
concept CELESTIAL OBJECT, connected to five nodes representing
attributes. By 1700 many mechanical philosophers had accepted that
astronomical objects move freely through space and that their orbits
play an important role in making predictions that can be checked
against observation. Hence our attribute list begins with the concept
ORBIT CENTER, followed by ORBIT SHAPE. Other attributes listed
here include DISTANCE, LUMINANCE (the source of the object's
light), and SIZE. All celestial objects possess all of these attributes.
Again, attributes are listed in a particular order purely for convenience;
the appearance of certain attributes at the top of the list does not indicate that they are more important than other attributes. The diagram
is labeled a partial frame because there are many other attributes that
might be included but are not listed here (nothing has been said about
physical constitution, for example).

Each attribute can take a number of distinct values. For example, empirical observation up to 1700 had shown that celestial objects may have orbits centered on stars or planets. It was not clear whether the stars themselves followed orbits centered on a particular object, so another possibility is free motion – an orbit without a center. Similarly, Newton had shown that orbit shapes in most cases are conic sections and the most important classes are designated as ELLIPSE (including circles) and HYPERBOLA (including parabolas). Corresponding to the possibility of free motion with no center we introduce the third possibility, labeled OTHER. The values for distance are perhaps less obvious. In principle the values for DISTANCE should be represented as an indefinitely large range of numbers. However, absolute distances had not been established, in the absence of a numerical value for the universal gravitational constant. Only relative distances were available, although it was clear that observers on the earth lived in a space structured somewhat as follows: in our immediate vicinity is the moon, which clearly moves around the earth. Slightly farther away but still in our immediate vicinity are the planets, which in the mechanical philosophy are regarded as moving around the sun. Stars are known to be at enormously greater distances, although these distances remain to be measured with any exactness. Comets pass between regions so distant that their parallax is indistinguishable from stars, and regions where they have a measurable parallax comparable with that for planets and the sun. For our purposes we may classify objects as NEAR if they are similar in distance to the moon, MEDIUM if they are similar in distance to planets, or FAR if they are similar in distance to the most distant objects known, the stars. Judgments of size were originally based on comparisons of luminosity. The advent of the telescope provided another means of estimating sizes of nearby objects, although it had been known from antiquity that the sun was the largest nearby object and stars were similar in size to the sun. The main difference introduced by the telescope was to classify comets initially as the same size as planets – before determination of the universal gravitational constant showed their masses were too small for that to be true. Beginning with the partial frame for CELESTIAL OBJECT circa 1700 (Figure 46), we see that the similarity classes STAR, PLANET, MOON, and COMET may be distinguished by similarity and difference among the features identified in the frame (Figure 47).

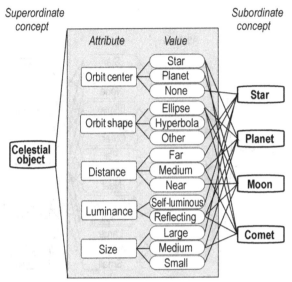

FIGURE 47. Partial frame for 'celestial object' in astronomy, circa 1700, showing subordinate concepts.

As we have already seen in Sections 4.3.1 and 4.3.3, the appearance of new similarity classes, that is, new subconcepts, will correspond to new sets of values distributed across various attributes, but this need not in itself create incommensurability. For example, comets may be divided into RETURNING and NONRETURNING without creating incommensurability between the earlier and later frames (Figure 48). In this case, the change is conservative: other entities are not redistributed and other classes are not redefined. Especially, no entity admitted in the new frame violates the no-overlap principle as it applies to categories in the old frame. All this is true, quite simply, because no new attribute-value combinations are introduced. The division of comets into two classes relies upon unexploited but accessible resources in the original frame.

The appearance of new attributes in a frame does not necessarily lead to incommensurability. We have already suggested that individuals may successfully identify the same equivalence classes by different relations of similarity and difference (Sections 2.3 and 3.3). When two such individuals compare notes, they may both add several nodes to their frames, but without changing their assignment of objects to

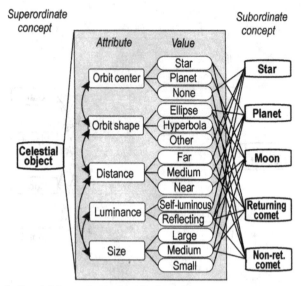

FIGURE 48. Partial frame for 'celestial object' in astronomy, circa 1700, showing new subconcepts 'returning comet' and 'nonreturning comet'.

equivalence classes, and hence without changing the overall structure of the taxonomy or frame. In this situation both individuals are using parts of a larger frame that is supported by their community. During a scientific revolution, however, taxonomies and frames change in a nonconservative way. Some entities are redistributed across categories (which are themselves redefined), while others appear and disappear. During the Copernican revolution, the meaning of the term 'planet' changed dramatically. One entity that had not previously been classified as a planet (the earth) now became a member of the redefined class. Two entities that had counted as planets before the revolution (the sun and moon) were moved to other classes. Some entirely new entities appeared (celestial comets) while others disappeared (terrestrial comets, and, at a slightly different level, celestial spheres).

The changes initiated by Copernicus redistributed objects among persisting equivalence classes like 'planet' and 'star', although the conceptual structure changed to permit the appearance of new classes such as 'moon', which now designates not the unique satellite of the earth, but satellites moving around any planet. Returning to Figure 46, let us consider some differences between that frame and a detailed

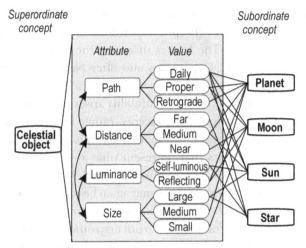

FIGURE 49. Partial frame for 'celestial object', circa 1500.

frame for CELESTIAL OBJECT in, say, 1500 before Copernicus had published anything (Figure 49). In 1500 CELESTIAL OBJECT already has the attributes SIZE and LUMINANCE. The question of DISTANCE had also been settled in a satisfactory way. But where there are two nodes for ORBIT CENTER and ORBIT SHAPE in the 1700 frame, the 1500 frame displays a single node, here labeled PATH. The path of a celestial object is its movement as viewed by an observer on the earth, and specifically its angular position from an agreed reference line. A planet's path has three components: its DAILY motion, its PROPER motion against the background of fixed stars, and (excepting the sun and moon) its occasional RETROGRADE motion. (The concept of path will be discussed in greater detail at the beginning of the next chapter.) The difference in the membership of the various classes of celestial object before Copernicus and (say) after Newton is striking. But incommensurability is generated here by the appearance (and disappearance) of entire attributes, and their associated values. Before Kepler, astronomical theories were concerned only to predict the angular position of a planet – no attempt was made to calculate what we would now call the orbit as a continuous track through space. Kepler was actually the first astronomer to attach physical significance to this track, at the same time that he introduced the modern concept of an orbit. (This case will be discussed in detail in the next chapter). Hence the top two attributes in Figures 46, 47, and 48 simply do not appear in the pre-Copernican frame for CELESTIAL OBJECT (Figure 49). The frames show that the concepts of CELES-TIAL OBJECT before Copernicus and after Newton are therefore incommensurable.

In the frame model incommensurability arises between conceptual structures, that is, patterns of concepts, rather than individual concepts. In its simplest terms incommensurability is a mismatch between the nodes of two frames that represent what appear to be the same superordinate concepts. Reading from left to right we may encounter the same series of attributes. These may again be represented by frames in a recursive manner. But at some point we encounter structures in the two series of recursive frames that do not map onto each other. Not just any mismatch will do. Division of subordinate concepts into further subclasses, as in the example of returning and nonreturning comets, relies upon unexploited but accessible resources in the original frame, usually unexploited value combinations. Hence, division of the super-ordinate concept that the frame represents into subclasses preserves the overall topology of the frame and will not generate incommensu-rability. The most serious problems will arise from the addition and deletion of attribute nodes. Incommensurability occurs between two frames for the same superordinate concept when we are confronted with two seemingly incompatible sets of attribute nodes. Unlike a tax-onomy, which shows only similarity classes, the frame representation makes explicit the attribute-value combinations that give rise to these classes. The frame notation therefore permits the direct representa-tion of incommensurability as a mismatch between frames, in a way that taxonomies alone do not.

In the end the differences that matter – and generate incommen-surability – are just those that create or reflect differences in attributes and values. These will be differences that correspond to differences in what Kuhn called similarity and dissimilarity classes, and hence in the fundamental objects that the conceptual structures represented by the frames allow us to talk about. And incommensurability is also a matter of degree – the higher the taxonomic level of the con-cept where the mismatch begins, the more severe will be the incom-mensurability. But there must be some connection between the two structures. Newtonian astronomy is incommensurable with Ptolemaic astronomy, but not with Galenic medicine. They are just two different fields.

An account as detailed and complex as Kuhn's, or the parallel account in terms of dynamic frame, offers many different ways of generating incommensurability. Kuhn did not explore all of these possibilities (Chen 1997, and Nersession and Andersen 1998 have gone some way beyond his account). In this chapter and the next we will be largely concerned with the simplest kind of incommensurability – mismatch between attribute nodes. However, we will not confine our attention to hierarchies of kinds but will consider quite generally the concepts needed to understand that portion of astronomy that deals with the planets during the sixteenth and seventeenth centuries.

5.4 GALILEO'S DISCOVERIES AND THE CONCEPTUAL STRUCTURE OF ASTRONOMY

At the time of Copernicus all natural philosophers in the Latin West agreed that the earth was the center of the cosmos, and that celestial objects somehow moved around it. The main physical constituents were a series of spherical shells, centered on the earth. Celestial objects like planets and stars were minor imperfections in these shells. They were carried around the heavens by the spheres as they moved (Swerdlow 1976; Van Helden 1985). How planets moved was a matter of bitter dispute. Averroist natural philosophers believed that the heavens consisted of a series of shells of the element ether, all concentric to the earth (Barker 1999). Ptolemaic astronomers agreed that overall the planets moved within a series of nesting concentric shells, but they gave a detailed account of the shells for each planet that included some parts generating circular motions not centered on the earth. In both cases the overall construction of the heavens was intended to conform to the principles of Aristotle's physics, and it was generally agreed that all celestial motions were compounded from motions that were circular and performed at constant speed (Barker and Goldstein 1998).

Although the fixed stars actually appear to follow paths across the sky that are circles traversed at constant speed, it is well known that the sun, moon, and planets do not. The planets are the most complex case, possessing both a proper motion in the opposite direction from

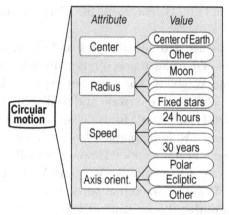

FIGURE 50. Partial frame for 'circular motion' as applied to astronomy, circa 1500.

the twenty-four-hour daily motion of the stars and a regularly repeated reversal of this motion called retrogression. Averroists and Ptolemaic astronomers differed radically in the explanations they gave for these two aspects of a planet's motion. Let us consider each of these positions in turn.

Figure 50 shows a partial frame for circular motion as it applies in understanding the motion of celestial objects. The concept has four important attributes for Averroists. First, all circular motions take place about some definable center, and for an Averroist this center must be the center of the cosmos (which is also the center of the earth). Although other centers of circular motion are geometrically possible, for physical and metaphysical reasons only one value of this attribute is allowed in any Averroist account of the heavens. Second, all circular motions must have a definite radius, although in practice Averroists were unable to specify precise values. It was generally recognized that for the heavens, the minimum radius was that of the motion of the moon – the nearest object – and the maximum was that of the fixed stars – assumed to be at equal distances and forming a boundary to the cosmos. In principle planets could move on circles at any radius between these boundaries. An array of boxes appears in the frame between MOON and FIXED STARS to indicate an indefinitely large range of intermediate values. The possible values for speed, or angular velocity, range from 24 hours – the speed of the daily rotation – through

FIGURE 51. Geocentric system of the world. Reproduced from Peter Apian, *Cosmographia*, Antwerp (1540), fol 6 R. Copyright the History of Science Collections, the University of Oklahoma Libraries, and reproduced by permission.

the slowest proper motion, that of the planet Saturn, which returns to its original position in the sky in slightly less than 30 years. Again, these values were seldom specified with any precision by Averroists, but the speeds of rotation for the proper motions of other planets must fall in the range between 24 hours and 30 years, also indicated by an array of intermediate boxes.

For the Averroists and their rivals, the physical constitution of the heavens consisted not so much of planets moving in circles as of planets carried by spheres. The celestial spheres were hollow shells that fitted perfectly inside one another (Figure 51). The sixteenth-century

name for a spherical shell bounded by two spherical surfaces is an
'orb'. Although astronomers made this technical distinction, they
often spoke of 'spheres', expecting their audience to understand that
they were referring to orbs.

In all cases the circles used in describing the motions of particular
planets are believed to result from the uniform rotation of an orb. For
an Averroist the direction of the orb's axis is an important variable –
here displayed as the fourth attribute node. While the axis for the
orb creating the daily rotation passes through the celestial poles, the
axis for the orb producing the proper motion coincided with the axis
of the ecliptic. Most importantly, retrogressions are created by the
combined effects of at least two concentric orbs with offset axes, carried within the orbs for the daily and proper motion (Pedersen 1993:
63–70, 235–236).

The Averroist account of the path for a planet was built from a
minimum of four circular motions, corresponding to four concentric
spheres (Figure 52). The spheres fit together perfectly, one inside
another, with no empty space between. It is assumed that the axis of
an inner sphere is carried by fixed points on the next sphere out.
Consequently, a planet carried on an inner sphere does not perform
a simple circle when viewed from the central earth. It follows a path
that is the resultant of the motions of the sphere that carries it and all
the spheres to which that one is attached, directly or indirectly.

The Averroist conceptual structure for the path of a planet may be
presented as a recursive frame diagram (Figure 53), using the frame
for circular motion (Figure 50) to specify the attributes and values
of each circular motion involved in the frame for PATH. Averroists
and Ptolemaic astronomers give identical accounts of the daily motion
(the fixed stars rotate about an axis through the poles once in twenty-
four hours, carrying everything else with them). To simplify our diagrams the corresponding branch will not be included in the next few
figures. Omitting the daily motion for simplicity, the Averroist conceptual structure will have one frame corresponding to the proper
motion and two corresponding to retrograde motions (as indicated
previously, one circular motion accounts for the proper motion, while
two circular motions account for retrograde motion). The recursive
frame diagram shown in Figure 53 incorporates a corresponding number of iterations of the frame for circular motion (Figure 50). Since

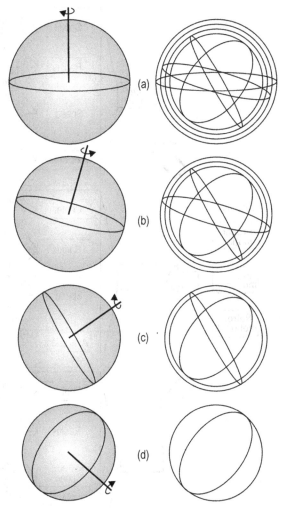

FIGURE 52. Averroist orb cluster, showing concentric orbs for daily motion (a), proper motion (b), and retrogression (c) and (d).

Averroists allow only a single value for the attribute CENTER, an identical node (CENTER OF EARTH) is activated in each circular motion considered here. Under PROPER MOTION, the value for AXIS ORIENTATION will be ECLIPTIC. Under RETROGRADE, the values will be neither POLAR nor ECLIPTIC, but specific to individual planets, here shown as OTHER.

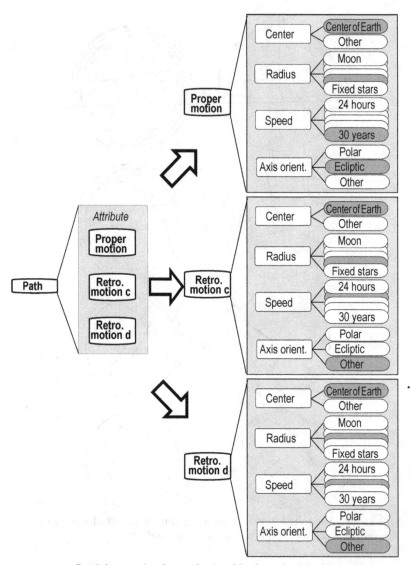

FIGURE 53. Partial recursive frame for 'path' of a celestial object, Averroist version.

The Ptolemaic account of a planet's path requires a simpler recursive frame and the activation of a different set of value nodes. The basic explanation for a planet's path, in addition to its daily motion, makes use of two mechanisms: an eccentric deferent and an epicycle

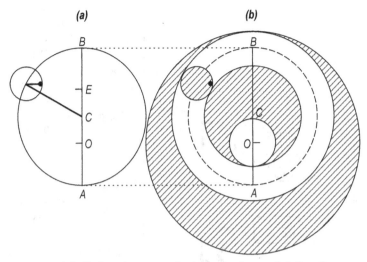

FIGURE 54. (a) Ptolemaic eccentric-plus-epicycle model for the proper motion and retrogressions of an outer planet. (b) Cross section through a set of Ptolemaic spherical shells that reproduce the circles of (a) as they rotate.

(Figure 54(a)). The complete model for planets like Mars, Jupiter, and Saturn makes use of an additional feature called the equant, which will be discussed in the next chapter. The two main features of the planet's motion, its proper motion and retrogressions, are explained primarily by the separate circular motions of the deferent and the epicycle, respectively (Pedersen 1993: 81–87).

There will be an obvious difference between the recursive frame for this part of Ptolemaic astronomy (Figure 55) and the corresponding Averroist frame (Figure 53). The lower part of the diagram, corresponding to RETROGRADE MOTION, will consist of two frames for CIRCULAR MOTION in the Averroist case but only one in the Ptolemaic case. This is not the kind of difference that creates incommensurability. As in the case of the comets discussed in Section 5.3, the greater complexity of the Averroist recursive frame is again created without introducing any new kinds of attributes, or new ranges of values. The two recursive frames do not, therefore, allow the appearance, in one frame, of objects that violate the no-overlap principle in the other.

In the Ptolemaic frame (Figure 55), while both the deferent and the epicycle correspond to circular motions, neither has the same

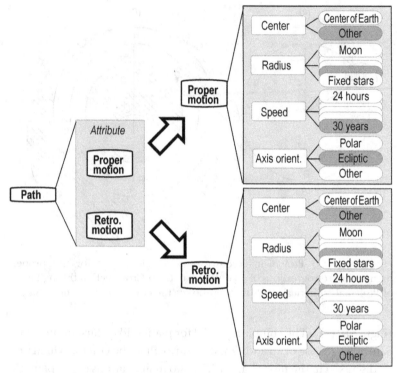

FIGURE 55. Partial recursive frame for 'path' of a celestial object, simple Ptole-maic version.

values assigned to attributes as in the Averroist case. The deferent is not centered on the earth but at a point some distance away and is therefore an eccentric circle. The epicycle center is carried on the eccentric deferent as it rotates, and its center is therefore remote from the center of the earth.

As did the Averroists, Ptolemaic astronomers believed that the circles in their planetary models were generated by the uniform rotation of earth-centered spheres. The eccentric deferent is generated by the rotation of two nonuniform spherical shells, or 'orbs', which appear as crescent shapes when displayed in cross section (see Figure 54(b)). Although the uniform gap between these shells is usually displayed in a contrasting color, it is itself a further solid object in which the small sphere representing the epicycle is physically embedded. The planet in turn is physically embedded in this minor sphere. With the exception of the epicycle sphere, all these shells rotate about axes that pass through the center of the earth, so that even the motion of the epicycle can be seen to be constrained by earth-centered spheres, which move in conformity with Aristotle's physics. This did not prevent Averroists from objecting to both eccentrics and epicycles on the grounds that the individual circular motions had impermissible centers. To decide between the Averroist position and the Ptolemaic position, the best evidence would be an example of a celestial motion that was inarguably centered at some other point than the center of the earth. This is exactly what Galileo provided.

Two pieces of telescopic evidence collected by Galileo between 1610 and 1613 could be used as decisive arguments against the Averroist conceptual structure. First, the observation of the phases of Venus seemed to require that Venus travel on a circle centered on the sun (Drake 1990: pt. 3). It is important to note that the *pattern* of the phases obliges this conclusion and not the mere observation of the phases themselves. Both Averroist and Ptolemaic accounts of the motion of Venus predict phases that appear as Venus moves away from the direct line between the earth and sun (Ariew 1987). However, in the Averroist account the fact that Venus never moves farther than about forty-six degrees away from this line would limit the observable phases to crescents, and the requirement that Venus be carried on a sphere concentric with the earth would make all phases the same apparent size. Galileo actually observed a full range of phases with widely varying sizes. In particular the (nearly) full phases were small, suggesting they took place on the far side of the sun, while the crescent phases were large, suggesting they took place nearer the earth. Although inconsistent with the original Ptolemaic account of the location of Venus, Galileo's results could be accommodated by the simple expedient of moving the center of Venus' epicycle from its original position on the earth-sun line, to coincide with the position of the sun. A Ptolemaic astronomer might well have said that Galileo's observations of phases for Venus confirmed the Ptolemaic account of its motion using an epicycle and accurately located the center of the epicycle for the first time (Ariew 1999: 97–119).

An even clearer case for non–earth-centered motion could be made from the discovery of Jupiter's satellites. In the very first book on his telescopic discoveries, Galileo (1610/1989) argued persuasively that

Jupiter was accompanied by four satellites moving on circles of different sizes around the planet as it traveled through the sky. The general acceptance of Galileo's discovery of these new objects made it
impossible to maintain the Averroist prohibition on centers of motion
other than the center of the earth. (Note, however, that neither of
these pieces of evidence in itself establishes whether the earth is in
motion around some external center, or vice versa.) Returning to the
frame diagram for circular motion (Figure 50), we may now summarize the dispute between the Averroists and the followers of Ptolemy
as follows: because the Averroists insisted that only one center was
allowed for celestial motions, they not only denied the possibility of
values other than their preferred value for the attribute CENTER, but
can be seen as rejecting the inclusion in the recursive frame of any
attribute-value pairs other than their preferred one. The Ptolemaic
astronomers, on the other hand, insisted that it was at least a legitimate question to inquire, for any particular motion used in astronomy, whether the center was identical to the center of the earth or
some other point, and their conceptual structure made use of some
of these additional value nodes. Galileo's telescopic discoveries vindicated the Ptolemaic insistence on the inclusion of these nodes, by showing that several celestial motions could not be accommodated without
them.

When Galileo's telescopic discoveries are analyzed in this way we can
see why the initial response to them did not lead to major changes in
the conceptual structure of astronomy. Many Ptolemaic astronomers,
for example, the Jesuits trained by Christopher Clavius at the Colle-
gio Romano, rapidly endorsed the telescopic discoveries (Lattis 1994).
Although the phases of Venus and the satellites of Jupiter require the
recognition that some value nodes for 'center of motion' must be
accepted beyond the Averroist choice, the corresponding attribute
node was not yet identical to the node appearing in the seventeenth-
century structure we examined earlier. Figure 46 contains nodes for
ORBIT CENTER and SHAPE. An orbit is a continuous track in space
traced by a planet, and it defines both the direction from the observer
to a planet and its distance. In all astronomical theories before Kepler
predictions were confined to directions, that is, angular positions of
planets with respect to a fixed reference line in space (Barker and

Goldstein 1994). The node for CENTER in the conceptual structure of Ptolemaic astronomy designates a center of angular motion and not an orbit center. After Kepler introduced the concept of an orbit in his 1609 *Astronomia Nova*, anyone accepting the new conceptual structure for astronomy presented there would be obliged to substitute a node that did represent ORBIT CENTER, together with a variety of choices for the shape of an orbit. Initially the two most important choices are the circle and the ellipse. Newton demonstrated that motions subject to an inverse-square law created orbits that were conic sections and may be seen as adding a new set of value nodes to an existing structure in which the attribute nodes were provided by Kepler and Galileo.

From the viewpoint of Kuhn's account of conceptual change in science, two points about this reconstruction deserve special mention. First, it can be seen that the transition from the Ptolemaic conceptual structure to the Newtonian one was not a process that took place instantaneously, but rather one in which an existing structure was successively modified. Kepler's theoretical work and Galileo's telescopic discoveries happened at almost the same moment. Kepler could argue, in favor of the new structure that he proposed, that by means of his new style of calculations he was able to specify the position of the planet Mars with an unprecedented accuracy. But the existence of the separate set of arguments, based on Galileo's discoveries of the phases of Venus and moons of Jupiter and supporting a conceptual structure diverging from the Averroist one in the same way as Kepler's, meant that his work and Galileo's rapidly became mutually supportive in the emergence of what was ultimately Newton's conceptual structure.

The same considerations also allow us to locate and appraise some of the most important incommensurabilities between pre-Newtonian and post-Newtonian astronomy. The first serious incommensurability appears with the replacement of the attribute nodes for PATH with those for ORBIT CENTER and ORBIT SHAPE. But it is important to recognize that other parts of the frame remained constant despite this change. Consequently the many astronomical questions that drew primarily on the attributes of celestial objects that remained unaffected by the change were uncontroversial, and supporters of both pre- and post-Newtonian astronomy could agree on their solution. As late as 1728 Ephraim Chambers found it useful to present the elements of both conceptual structures in a single work. The idea that the introduction of any change in a conceptual structure leads to total communication failure between supporters of the new structure and supporters of the old is therefore seen to be completely unfounded.

In Section 4.5 we suggested that conceptual changes vary in degree, and that the frame account explains why some changes are more severe than others. By the same token, the degree of severity of incommensurability may also be appraised by analyzing where replacements are made in the frame. Roughly speaking, as described in Section 4.5, the higher in a kind hierarchy the replacement of attributes appears, the more acute the problem will be. Phrasing the point in terms of frames, if we consider the frame for a multiple-level kind hierarchy (Figure 16) we can say that the more general the attribute that is replaced, the greater the incommensurability. The mark of incommensurability between two conceptual structures is therefore not a total failure of correspondence between them, but rather the appearance of two or more attributes that differ and that introduce different sets of values. In general, merely introducing a new set of values for an existing attribute will not generate incommensurability. Averroist astronomy and the simple version of Ptolemaic astronomy we have discussed so far are not incommensurable, although the full version may be (as we will see in the next chapter). The addition or deletion of an attribute will create incommensurability only if the new attribute-value sets violate the no-overlap principle (or another of the hierarchical principles introduced in Section 4.2) as applied to the attribute-value sets of the previous frame. Thus, the incommensurability between Keplerian astronomy and Ptolemaic astronomy created by the deletion of the attribute PATH in favor of the attributes ORBIT CENTER and ORBIT SHAPE is significant (Figures 46 and 49), but the incommensurability between the concept of PHYSICAL OBJECT in post-Newtonian physics and in pre-Newtonian physics will be considerably more severe, as that concept is superordinate to the concept of an astronomical object which we have been considering (Figures 43 and 45).

The difficulties labeled incommensurability have so far appeared when two or more conceptual structures from different scientific traditions have been compared. We will see in the next chapter that individual traditions may suffer from similar difficulties. Copernicus' main announced objection to Ptolemaic astronomy may be seen as a problem of just this kind.

6

The Copernican Revolution

6.1 THE CONCEPTUAL STRUCTURE OF
PTOLEMAIC ASTRONOMY

In astronomy before Kepler the path of a planet was not its orbit but the pattern of its motion seen by an observer on a stationary earth against the hypothetical sphere of the heavens. It was recognized in antiquity that this pattern was not a real motion, but a complex outcome of the observer's viewpoint and a variety of circular motions that acted together. The task of astronomy was to define this pattern – to specify the path of the planet in this original sense. The real motion of the planet – its track in what we would now call three-dimensional space – was unknown, and possibly irrelevant. All other considerations – the causes of celestial motion, the actual dimensions of the heavens – were the business of a separate science, cosmology. It was well known that the goal of astronomy could be achieved, that is, the path of a celestial object could be predicted, without making specific assumptions about its distance from the earth, once appropriate rates of rotation were introduced (Evans 1998; Pedersen 1993).

The basic data of astronomy from antiquity to the sixteenth century – the *explananda* or, if you prefer, the 'phenomena' that needed to be 'saved' – were recorded observations of planetary positions. Sixteenth-century astronomy texts devoted most of their attention to motion in longitude. Motion in latitude was usually handled by a brief section at the end of the book and after the main business had been

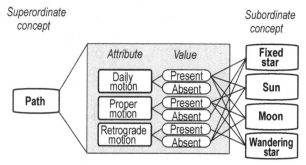

FIGURE 56. Partial frame for 'path', circa 1500.

completed (Evans 1998). It is also worth noting that the data to be explained were extremely sparse: before the programs of systematic observation initiated at Kassel by the Landgrave Wilhelm IV and at Hven by Tycho Brahe, most observations of celestial positions were made when celestial objects were doing something unusual, such as retrogressing or passing close by another object (Thoren 1992; Hamel 1998; Christianson 1999).

The details of the planets' real motions remained controversial. Averroists insisted that the real motion must be concentric to the earth (Section 5.4). Ptolemaic astronomers insisted that it must be at least eccentric and modified the eccentric motion with an epicycle in all cases except the sun. The shift from Ptolemy to Copernicus and Kepler therefore includes a conceptual shift from a concept of path that is not an orbit and does not specify a real motion, to a concept that specifies an orbit and is a real motion. To understand the first of these changes in greater depth we return to a detailed consideration of changes in the frame for CELESTIAL OBJECT from before Copernicus to after Newton. Rather than rewriting the whole frame for CELESTIAL OBJECT we will diagram only the attribute PATH and its values (Figure 56).

As already indicated, PATH does not refer to the continuous motion of an object through three-dimensional space. It refers rather to the pattern of an object's motion viewed against the sphere of the heavens. Kuhn expressed this well in *The Copernican Revolution* when he referred to the 'two sphere' universe – the central earth surrounded by a hypothetical sphere of the heavens (Kuhn 1957: Ch. 1). It is the goal of astronomical calculation to calculate the successive positions of the

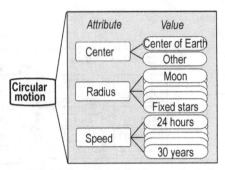

FIGURE 57. Partial frame for 'circular motion', circa 1500.

planet along this path – including any retrogressions. As the size of the sphere of the heavens is arbitrary, the only important data here concern changes in angular position. It is not part of the main business of astronomy to calculate planetary distances, although this may be done in an ancillary set of calculations that appeal to some premises outside astronomy (for details see van Helden 1985; Barker and Goldstein 1994). It is, however, generally acknowledged that the moon is the closest celestial object, and that the fixed stars are the farthest. The irrelevance of distance in astronomy becomes clearer if we reconsider the frame for a related concept: CIRCULAR MOTION (Figure 57).

To make the frame more compact Figure 57 uses a simplified version of Figure 50, omitting the node for AXIS ORIENTATION, which adds nothing important to this phase of the analysis. Circular motions have three remaining attributes that need to be considered here: they have a CENTER, a RADIUS, and a rate, which may be designated by an angular velocity or SPEED. As far as astronomy is concerned, the angular velocity can again take any numerical value, so for constant angular velocities (uniform motions) the value nodes connected to this attribute should be an infinite array, limited by the known maximum (one rotation in 24 hours) and minimum (Saturn's rotation in just under 30 years). In principle a circular motion must have a definite radius. In practice this plays no role in astronomical calculations; however, we may display the range of possible values by an array of boxes, starting with the minimum radius (the distance to the moon) and ending with the maximum (the distance to the fixed stars). The possible values of the attribute CENTER are also indefinitely many. But again we may simplify matters by picking the single most important

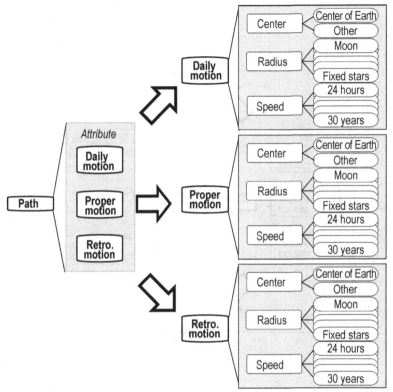

value (the center of the cosmos, which is also the center of the earth) and grouping all the rest as OTHER. Using this frame, the main difference between the Ptolemaic tradition and its main rival during the sixteenth century reduces to the activation of a single value node. Ptolemaic astronomers allow the activation of the OTHER value node for CENTER; their opponents the Averroists do not.

All celestial motions are circles traversed at constant speed. This fundamental tenet of sixteenth-century astronomy can be displayed by combining the frames for PATH and CIRCULAR MOTION circa 1500 (Figure 58). PATH has three attribute nodes. Each of these is a separate motion; hence each must display the attributes already introduced for CIRCULAR MOTION. Using the recursive property of frames each attribute in the frame for PATH can be expanded into a new frame with its own attributes and values. Ptolemaic astronomers use a single

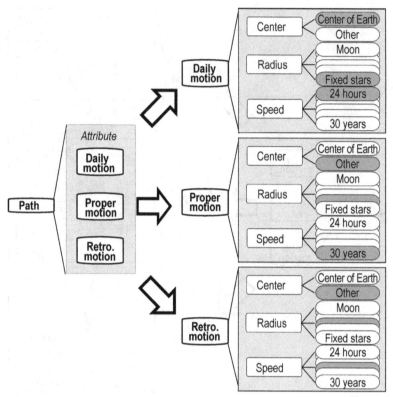

FIGURE 59. Partial recursive frame for 'path', circa 1500, with activated value nodes representing a Ptolemaic theorica for Saturn.

model, called a *theorica* (plural *theoricae*), with varying numerical values, to represent the motions of the outer planets and Venus. By activating specific value nodes in Figure 58 we can generate the conceptual structure of a Ptolemaic theorica for the planets Venus, Mars, Jupiter, and Saturn, which were usually treated together (Figure 59). Individual theoricae for different planets will differ primarily in the numerical values assigned for the attributes CENTER and SPEED. As already indicated, the attribute RADIUS plays no role in determining PATH. If specific values are assigned, they are provided by cosmological not astronomical reasoning, and all values for planets fall within the range that takes the distance to the moon as its lower limit and the distance to the fixed stars as its maximum. Saturn is accepted to be the most distant planet. To make our analysis more concrete, typical values for Saturn will be used in the subsequent frames. Note that the diurnal motion is earth-centered, but that the proper motion and the retrograde motion have different centers, corresponding to the center of the eccentric deferent and the epicycle, respectively. The complication caused by the notorious equant point is deferred for later discussion (Section 6.3).

Other theoricae are simple variations on this basic pattern. The theorica for the sun does not contain the attribute RETROGRADE MOTION in the frame for PATH and has different values for the attributes under proper motion. That for the moon has the attribute-value sets representing a nonretrograding epicycle as its third element. Next let us consider how this frame changes to accommodate the astronomical models proposed by Copernicus in 1543.

6.2 THE CONCEPTUAL STRUCTURE OF COPERNICAN ASTRONOMY

Figure 60 is a partial recursive frame for PATH circa 1543, with activated value nodes representing a Copernican theorica for Saturn. The frame for DAILY MOTION now represents the twenty-four-hour motion of the earth, which for Copernicus includes everything inside the sphere of the moon. Copernicus' treatment of the diurnal motion uses practically the same attribute-value combinations as Ptolemy's. Only the radius of the motion changes from the largest allowed value to the smallest. The most important change, of course, is that this whole motion is now regarded as a real motion of the earth, not the fixed stars. When it comes to making astronomical calculations this makes no difference at all.

The proper motion of each planet is now understood primarily through the motion of a large circle eccentric to the mean sun. Notice that this change requires *no* addition or deletion of attributes, and no major differences in the activated values. Within the conceptual system of sixteenth-century astronomy, the choice of the mean sun as center for the proper motion is just the choice of a new center for the eccentric circle that differs from the center of the earth. But Ptolemaic astronomers were already using such points in all their models.

Turning to the other two value nodes in the frame for PROPER MOTION: the distances involved remain comparable to the Ptolemaic

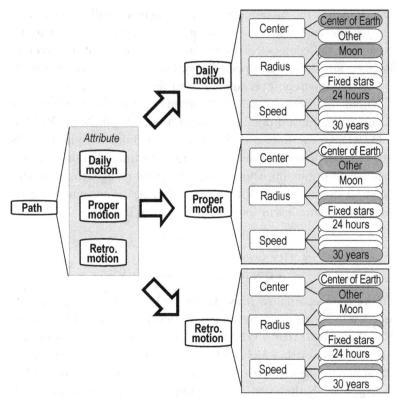

FIGURE 60. Partial recursive frame for 'path', circa 1543, with activated value nodes representing a Copernican theorica for Saturn.

ones, although for Copernicus particular values generally become larger than their Ptolemaic equivalents. Although there is now a relation among these distances (changing one requires that you change the rest) this connection has no consequences for calculating planetary positions. The SPEED of these motions also assumes different (but related) values from those in the Ptolemaic frame, but again, these are drawn from the existing set of allowed values. So in the cases of all three attributes we activate values within ranges already admitted in Ptolemaic astronomy.

What has just been said describes the mathematical models for calculating planetary positions presented in the body of *De Revolutionibus*, and not the cosmological sketch from Book I. Once again, in the case of Copernicus' account of proper motion, the major difference from

Ptolemy is obscured if we concentrate only on the basis for astronomical calculations. For Ptolemy, the main element in the proper motion is an eccentric circle. But exactly the same results would follow from a concentric circle carrying an epicycle, and the absolute size of these circles is arbitrary. So judging only from its path we cannot discern the real motion of the planet. For Copernicus, however, the motion that results from the eccentric circle and its ancillary minor epicycle is a real motion, with a definite spatial location.

Finally, let us examine the lowest of the three right-hand frames in Figure 60. In Ptolemaic astronomy (Figure 59) this frame represents the properties of the epicycle used to accommodate retrograde motion. For Copernicus, retrograde motion is the result of the annual motion of the earth combined with the proper motion of the planets that has already been introduced. As is well known, Copernicus explains retrogression through change in the line-of-sight as a moving earth overtakes an outer planet or is overtaken by an inner one (Kuhn 1957:166–169). At the same time he can explain why outer planets retrogress while in opposition to the sun (and inner ones in conjunction), and why the retrogressions begin and end where they do (Kuhn 1957:165–167).

To derive actual positions for retrogressions we need a theorica for the earth to replace the theorica for the sun. Copernicus provides this by giving the earth a purely circular path centered on the mean sun. So, just as Ptolemy does, he introduces a circular motion to explain retrogressions. The attributes and values of this circular motion are surprisingly familiar. Of course the center of this motion is the mean sun, that is, a hypothetical point differing from the center of the earth. The speed of this motion is the speed attributed to the sun in the Ptolemaic theorica. And the radius of this motion is the distance attributed to the sun in the Ptolemaic theorica. Again Copernicus' theory introduces no new attributes, and the values he uses for the attributes already introduced fall within the ranges already admitted in Ptolemaic astronomy.

In our reconstruction Copernicus uses the same overall structure as Ptolemy for the key concept PATH, which encompasses the positional data of astronomy. We do not need to add attributes or values, we do not need to delete attributes or values, and we do not need to add new *kinds* of attributes or values. Not only does Copernicus employ the same attributes, but the values activated in his frame are almost the same pattern as the Ptolemaic ones (in contrast to those activated in an Averroist account) and there is nothing objectionable in the particular values assigned to these attributes. This includes the attribute-value sets used in the treatment of retrogression, in which the proper motion of the earth serves the same function as the epicycle in a Ptolemaic theorica. So if incommensurability is judged by degree of mismatch between attribute and value nodes, the conceptual structures of Ptolemaic planetary astronomy and Copernican planetary astronomy (Figures 59 and 60) are not incommensurable. Now remember that it is the goal of astronomy in the sixteenth century to calculate planetary positions against the sphere of the heavens as viewed from the earth. The astronomer reading *De Revolutionibus* is reading it with that goal in mind. And with that goal in mind the sixteenth-century astronomer will find the conceptual structure underlying Copernican calculational techniques to be the same structure that appears in Ptolemaic astronomy. Copernicus' intent is to restore an astronomy that uses only the attributes of circular motion that we have displayed in our frame, and hence to conserve an existing conceptual structure.

Our analysis has concluded that Copernicus' conceptual structure is not incommensurable with Ptolemy's – if anything, it appears to be a variation on it. This is exactly the way Copernicus and Ptolemy were seen during the late sixteenth and early seventeenth centuries. Erasmus Reinhold at the University of Wittenberg adopted Copernican calculation techniques to produce a new and improved set of astronomical tables (the *Prutenic Tables*). Following his lead, a whole host of Ptolemaic astronomers spread Copernican methods through Northern Europe. Farther south, the Jesuit Christopher Clavius, who led the successful reform of the calendar, also counted Copernicus as an intellectual ally of Ptolemy in the common fight against the Averroists (Gingerich 1993; Lattis 1994; Barker and Goldstein 1998).

6.3 THE PROBLEM OF THE EQUANT POINT

In the previous section we established that the Copernican account of celestial motions and the simplest Ptolemaic account use the same conceptual structure. The same result was established in the previous chapter for the Averroist account and the Ptolemaic account.

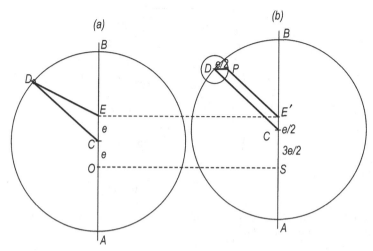

FIGURE 61. (a) Ptolemaic eccentric-plus-equant model for an outer planet, compared with (b) Copernican model with 'concealed equant' at E'.

According to the simplified frame introduced in the current chapter, the conceptual structures used by Averroists and followers of Ptolemy differ only in the values assigned to a single attribute (CENTER) and whether certain values are allowed for that attribute. The introduction of the equant changes the situation (Figure 61).

Ptolemy probably recognized that the simple eccentric-plus-epicycle model failed to predict both the direction and the angular width of planetary retrogressions (Evans 1998: 355–359). To correct this he introduced a new device (see Figure 61(a)). Ignoring the epicycle for the present, and considering a diameter of the eccentric (line *AB*) that passes through the position of the earth *O* and the eccentric center *C*, Ptolemy defined a point *E* at the same distance *e* as the earth *O* from the center but on the opposite side. He then used this point *E*, which he called the equant, to control the motion of the epicycle that carried the planet (compare Figure 54(a)). In his complete model for outer planets, the center of the epicycle moves uniformly along the eccentric not when viewed from the geometrical center of the eccentric *C*, but when viewed from the equant *E*. By means of this subsidiary device Ptolemy was able to bring his theory into excellent agreement with observations based on the naked eye. However, from the viewpoint of conceptual structure, and the physical underpinnings of astronomy, this success in calculation is achieved at a very high price.

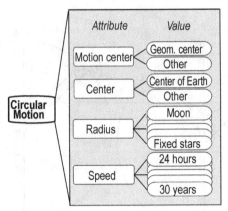

FIGURE 62. Partial frame for 'circular motion' showing modifications required to accommodate Ptolemaic equant.

In all previous frame diagrams for circular motions it has been taken for granted that the center used to define the radius of a motion and the center used to define the speed of a motion are the same point. In Ptolemy's complete model for the outer planets these are not a single point. In order to accommodate the equant, we therefore need to add a new attribute node to the basic structure for circular motion (Figure 62; compare Figure 57). It is not obvious how this revision should be made in the frames representing astronomical theories that we have considered so far (especially Figure 59). Although Ptolemy makes use of the equant only for a single one of the circular motions making up a planet's path, this change raises the question whether a similar revision is needed in the case of the other motions. Rephrasing this in terms of frame diagrams, the issue is whether to add a new attribute node only in the case of the circular motion corresponding to the eccentric that carries the epicycle (the proper motion) or in all the circular motions needed to specify the planet's path. In these other cases, and especially in the case of the epicycle used for retrogressions, it seems the value of the new attribute happens to coincide with the value for the circle's geometrical center (Figure 63; compare also Figure 59).

Adding an attribute node for MOTION CENTER is not a conservative revision of the prior conceptual structure. We may now recognize two classes of circles required in Ptolemy's theorica for outer planets and generated as subconcepts by the revised frame: circles in which

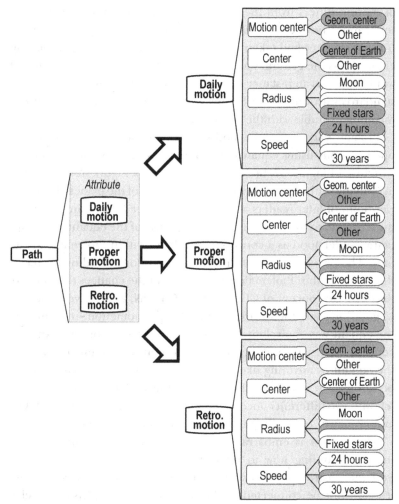

FIGURE 63. Partial recursive frame for 'path', circa 1500, with activated value nodes representing a Ptolemaic theorica for Saturn, showing modifications required to accommodate Ptolemaic equant.

the center of motion is identical to the geometrical center and circles in which it is not. When the new attribute node is introduced in the frame for CIRCULAR MOTION, an existing entity, the major circle that carries the epicycle, is reclassified from an existing category (circle for which the geometrical center and center of motion coincide) to a new and previously nonexistent category (circle in which the center of motion differs from the geometrical center). Entities of this new sort cannot be accommodated within the old conceptual structure. If these changes had taken place over time, with Figure 62 replacing Figure 57 as the generally accepted frame for circular motion, then it would count as an instance of revolutionary change. According to the standards introduced in Chapter 5, the later conceptual structure is incommensurable with the earlier one.

Ptolemy's account of the motions of the sun and the moon is developed without using equants, and hence using only the conceptual structure of Figure 57, but all of the features of Figure 62 are present in Ptolemy's subsequent account of the motion of the planets. From the viewpoint of later readers all these models date from a single source, the *Almagest*, so the difficulty of making sense of the equant is perhaps better understood as a conceptual problem within Ptolemaic astronomy. The peculiar status of the equant was a long-standing source of discontent within Ptolemaic astronomy, and the changes that it introduces in the concept of CIRCULAR MOTION, as we have presented it, go a long way toward explaining this phenomenon: resistance to the equant was equivalent to resistance to a revolutionary change in conceptual structure. Ptolemy does not motivate the introduction of the equant by anything like the specification of an anomaly that can be resolved by modifying the frame, so the student of the *Almagest* is left with two different conceptual structures for CIRCULAR MOTION and no way of reconciling the discrepancies between them.

The equant is embarrassing not only because of the difficulty in understanding how to revise the basic conceptual structure of Ptolemaic astronomy in order to accommodate it, but also because it could not be connected in the usual way with a physical mechanism. As already described, all other circular motions in Ptolemaic astronomy could be imagined as the result of uniform rotations of earth-centered orbs, or spheres carried by such orbs (Figure 54(b)). The equant motion could not be replaced by an earth-centered orb and could not be modeled by a uniform rotation of any of the orbs already accepted (Barker 1990).

Although Averroist natural philosophers objected to eccentrics and epicycles, the main difficulty that concerned Ptolemaic astronomers within their own tradition was the equant. The seeming impossibility of accommodating this necessary technical device within the basic

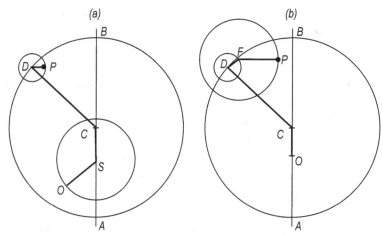

FIGURE 64. (a) Copernican model for an outer planet, heliocentric arrangement, compared with (b) Copernican model for an outer planet, geocentric arrangement.

conceptual structure of circular motion, or of connecting it with physical models in the usual way, led to the creation of an entire school in Islamic astronomy centered at Maragha in Persia, which developed new mathematical devices and equivalent systems of orbs to avoid it (Ragep 1993). By the fourteenth century this school had found a number of different, calculationally adequate means of avoiding the equant (Pedersen 1993: 241ff.). Although these results remained generally unknown in the West, it is clear that Copernicus encountered some version of them during his education in Italy (di Bono 1995; Barker 1999). When he published his new astronomical models, he was mistakenly given credit for many innovations that had actually occurred in Islam.

Copernicus' model for the outer planets avoids using an equant by adopting a device introduced by Ibn ash-Shātir of Damascus (1304–1376) (Pedersen 1993: 242–245). No new center of motion for points on the eccentric is introduced. Instead a small subsidiary epicycle is inserted in the model at point D (Figure 61(b) cf. Figure 64(a)). In Ptolemy's original model the distance from the equant to the center of the eccentric and from the center of the eccentric to the observer had been equal ($EC = CO$, in Figure 61(a)). Take the sum of these two distances to define a unit distance. Copernicus' model in effect retains the same magnitude for this total distance. He then assigns a distance of three-quarters of the unit between the center of the eccentric and the physical center of the system S (formerly the position of the observer on the earth, now the mean sun). Copernicus' model adds a minor epicycle carried by the eccentric at point D, the radius of which is the remaining one-quarter of the unit distance. In Copernicus' presentation this epicycle carries the planet. Its center moves uniformly about the geometrical center of the eccentric. However, the conditions placed on the motion of the minor epicycle (angle $BCD =$ angle CDP) are such that the planet carried by it moves uniformly with respect to a point E' farther along the center line AB (the line of apses) from the eccentric center (shown in Figure 61(b)). So, although the equant point does not appear in Copernicus' diagrams, it is still possible to define an equant point in Copernicus' models, and the planets move in just the way they would if their motion were controlled by an equant in the Ptolemaic manner (for a discussion see Evans 1998: 421–422; Voelkel 2001: 19). The ease with which an equant point can be defined for Copernicus' construction has led some modern commentators to question whether he eliminated the equant at all (Neugebauer 1968). These mathematical considerations should not, however, make us lose sight of more fundamental points about the conceptual structure of Copernican astronomy and its physical interpretation.

The motions described so far represent the main motion of the planet – its proper motion – around the mean sun. For Copernicus, retrogressions are explained by viewing the motion so defined from the moving earth (point O in Figure 64(a)), which is itself in motion around the mean sun S. For Ptolemy, the proper motion is described by the eccentric, while retrogressions are accommodated by the epicycle. In Copernicus' models, the motion of the earth around the sun, which is still treated as a circular motion, replaces the Ptolemaic epicycle. To make a prediction about the angular position of a planet in the sky, however, we still require not only the eccentric, but also this second circle or epicycle, in addition to the new minor epicycle that Copernicus has inserted as part of his mechanism to avoid using an equant point. Copernicus' model, then, can be represented as a double-epicycle system (Figure 64(b)). If the earth is placed at O, this converts the model back to a geocentric system, an option used by the group led by Erasmus Reinhold and now called the Wittenberg astronomers (Westman 1974; Barker and Goldstein 1998).

If the mean sun is placed at O in Figure 64(b), a double-epicycle heliocentric system appears. Kepler, for example, presents Copernicus in this way (Kepler 1609: 14).

In Copernicus' model, all the motions are simple circular motions that can be understood in terms of the original conceptual structure for circular motion presented in Figure 57. No separation of centers of motion from geometric centers is required. Second, because only simple circular motions are used, either Copernicus' original models or their geocentric equivalents can be represented by sets of orbs with centers either at the mean sun, in the case of Copernicus, or at the earth, in the case of Ptolemaic astronomers. In fact, the physical location of the center of the system is irrelevant to the success of the model as a calculating device. (The equivalence is easily seen in vector diagrams – see Figures 64(a) and 64(b).) So although Copernicus' model can be readily reinterpreted in terms of an equant, and although he is describing a motion that is originally defined by means of one, his real achievement is to specify a mechanism that avoids both the deviant conceptual structure required by Ptolemy's complete model and the associated problems of physical interpretation. This was clearly the response of his contemporaries, who regarded Copernicus as amending and improving Ptolemaic astronomy, rather than undermining it. Erasmus Reinhold wrote on the front page of his personal copy of Copernicus' book, "The first axiom of astronomy – all motion is in circles at constant speed" (Gingerich 1993). Georg Rheticus in his preliminary survey of Copernicus' theories simply announced that Copernicus had eliminated the equant (Rheticus 1540/1979: 136–137). And later thinkers like Maestlin and Kepler presented Copernican models that were consistent with this understanding of his work and that could be interpreted in terms of three-dimensional orbs (Kepler 1596).

If we compare the frame diagrams for the simple Ptolemaic model for the outer planets, Copernicus' model, and Ptolemy's full model including the equant (Figures 59, 60, and 63), it is apparent that it is Ptolemy's full model (Figure 63) that differs most from the other two, because it includes new attribute nodes in all the frames used to recursively expand the attributes of the superordinate concept PATH. If the addition or deletion of attribute nodes leads to the redistribution of entities across categories in ways that are prohibited in the unrevised structure, then the result is incommensurability. On this basis it can be said that Ptolemy's complete model is incommensurable with both the simple model and with Copernicus' model.

Although the introduction of the equant did not cause a failure of communication or lead to the impossibility of comparison between the full Ptolemaic model and alternatives, it can be seen from this reconstruction that the long-standing discomfort with the equant was motivated by a discrepancy in conceptual structures of exactly the same kind that we have already identified in cases of incommensurability between different scientific traditions. It should also be apparent that Copernican astronomy is not incommensurable with either the conceptual structure favored by the Averroists or the Ptolemaic alternative, apart from the difficulties with the equant, which the Ptolemaic astronomers regarded Copernicus as having resolved. But Copernican cosmology, with its central sun and the earth reclassified as a planet, is clearly incommensurable with Ptolemaic cosmology, with its central earth. How was this conflict avoided? The Wittenberg interpretation of Copernicus simply disregarded the cosmology (as obviously wrong on physical and scriptural grounds) and referred the astronomy to a central earth, using a model like that shown in Figure 64(b). This was the most influential interpretation of *De Revolutionibus*, from the death of Copernicus in 1543 until the appearance of major works by Kepler and Galileo in 1609 and 1610 (Westman 1975; Barker 2002). Kepler insisted on introducing physical considerations based on heliocentrism that led to a revision in the conceptual structure of astronomy and the first major incommensurability with the structures used by Ptolemy and Copernicus. However, to explain how this change came about we need to consider issues outside positional astronomy, and the conceptual structures we have considered so far.

6.4 FROM ORBS TO ORBITS

The obvious incommensurability between pre- and post-Copernican astronomy is indicated by the redistribution of existing entities across categories in ways forbidden by the pre-Copernican conceptual structure. There are two conspicuous examples. First, the sun moves from a category that includes the moon and planets to a new and special category (perhaps shared with the fixed stars, if they possess planetary systems). Second, the earth becomes a planet, a claim that is unintelligible in the previous conceptual structure (see Section 4.2.3). These are clear characteristics of the changes we have identified as revolutionary. Other features of such changes are the appearance of new entities and the disappearance of old ones. As we have seen in the previous section, these changes are not required by modifications in the conceptual structure of astronomy as it was understood in the sixteenth century. Rather these changes come about because of collateral changes that undermine an account of fundamental entities that applies in physics, cosmology, and, by extension, astronomy. These collateral changes undermine the status of the orbs that hold the sun, moon and planets and produce their motions. In a sense these are the real fundamental entities in sixteenth-century astronomy.

Before Copernicus it was agreed by almost everyone that, beginning with the region of the moon, the cosmos consists of a series of concentric shells with a planet somehow confined to each one (Figure 51). Averroists believed that the detailed motions of each planet could be recovered by dissecting each shell into a series of thinner concentric shells, with offset axes and varying rates of rotation, as described in Section 5.4 (see especially Figure 52). Throughout Copernicus' lifetime, systems of this sort were regularly proposed by natural philosophers who objected to Ptolemy's use of more than one center of rotation. The most detailed efforts are due to Giovanni Battista Amico in 1536 and Girolamo Fracastoro in 1538 (di Bono 1995). Ptolemaic astronomers, however, believed that the interior of each celestial shell is divided in another way. Although the inner and outer surfaces remain concentric to the earth, interior surfaces may be eccentric, creating shells, or orbs, as they are called in a theorica, that vary in thickness. These appear as crescent shapes when displayed in cross section (Figure 54 (b)). Two of these shells can be arranged to sandwich a third, which carries a small sphere corresponding to the planet's epicycle, or, in the case of the sun, the spherical body of the sun itself. As they rotate, these shells generate motions equivalent to the eccentric and epicycle. Where the cluster of shells for one planet ends, the next begins (Figure 65). By arranging systems of orbs inside each other, as illustrated here, a complete system of the world could be constructed, following the overall pattern of Figure 51. The planet is literally embedded in the epicycle sphere – indeed it is a commonplace

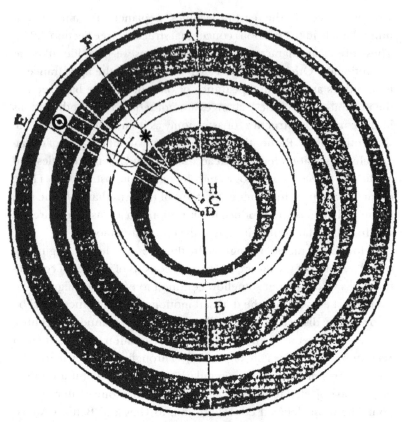

FIGURE 65. Combined orb diagrams for the theoricae of Venus and the sun. Erasmus Reinhold, *Theoricae novae planetarum,* Wittenberg (1542), fol. M viii R. The body of the sun ☉ is carried in the orb labeled A shown here in cross section. Inside the set of orbs including A, a further set carries the planet Venus ✳ on a small sphere that rolls inside the orb labeled B. The central Earth occupies point D. Copyright the History of Science Collections, the University of Oklahoma Libraries, and reproduced by permission.

that a planet is distinguished from the material of the orbs only by its density: "A planet is a denser part of its orb." So it might be said that the fundamental entities in sixteenth-century astronomy are the invisible celestial spheres and the orbs into which they are divided.

The celestial ontology of the sixteenth century had several problems. Averroists had the problem that no system of homocentric spheres had ever been demonstrated to produce the observed path of a planet. The Ptolemaic astronomers had the problem that the equant motion cannot be accommodated as a uniform rotation within such a shell cluster, and the further problem that Apollonius' theorem establishing the equivalence of eccentrics and concentrics carrying epicycles makes it possible to generate the same path in several different ways, each corresponding to a different arrangement of orbs. Thus Ptolemaic astronomers were in the dubious position of being committed to orb clusters but not being able to specify the actual pattern of orbs in the heavens. Careful people like Phillip Melanchthon confined themselves to the claim that some such cluster of orbs is up there (Melanchthon 1549/1846: col. 244). Other defenders of mathematical astronomy, such as Christopher Clavius, flatly insist on the existence of the specific clusters presented in his Ptolemaic theoricae (Lattis 1994; Barker and Goldstein 1998).

It is rather difficult to say what kind of substance these spheres are made of. Although they are conventionally referred to as 'crystalline' (Kuhn 1957: 79–82), this should be understood to indicate their transparency, and not their hardness. Hardness may have been attributed to them only later, and most conspicuously by people like Tycho Brahe who opposed solid spheres in favor of fluid heavens (for a discussion see Goldstein and Barker 1995).

Copernicus notes the general liabilities of both these positions in the letter to Pope Paul III at the beginning of *De Revolutionibus*. But little or nothing changes with Copernicus. As a perceptive Lutheran put it in 1576, you can generate the Copernican cosmos from the conventional one by swapping the sun and the earth-moon combination (Barker and Goldstein 1998). Everything else stays in place. Copernicus' cosmological diagrams are not pictures of orbits, but patterns of orbs. There is both internal and external evidence for this: the significance of his drawings is quite clear once you look for orbs in place of orbits (Swerdlow 1976; Barker 1990). Copernicus also continues to speak of planets being carried by their orbs or spheres. And Maestlin and Kepler, in their own presentations at the end of the century, clearly understood him as continuing to accept a nested shell cosmos (Barker 2000).

For Copernicus, each planet is again confined within a specific shell. How the shells are further divided is not specified – but it does not need to be. Anyone familiar with the orb models presented in a Ptolemaic theorica can construct equivalent patterns for Copernicus' detailed planetary models. To succeed at the business of astronomy – recovering planetary paths – Copernicus need not say anything new about the substance of the heavens, and he does not. And until the substance of the heavens changes, planets remain as minor flaws in much larger sets of divided spherical shells.

The arguments about the substance of the heavens and the demise of celestial orbs have several strands. One in particular concerns us here. Observers all across Europe described a comet that appeared in 1577. Many of them concluded that the comet was above the moon and thus a celestial object, contrary to Aristotle's teachings. Two observers were also unique in offering a new kind of data. They were Michael Maestlin, soon to become a professor at Tübingen and teacher of Johann Kepler, and Tycho Brahe, already beginning to establish himself on the island of Hven. Maestlin's account appeared immediately – Brahe's not for ten years.

Maestlin (1578) and Brahe (1588) published tables that gave the comet's position, that is, its direction, every day over a period of months. But they also calculated the distance of the comet from the earth for each position they gave. This was the first time anyone had described the continuous track of a celestial object as it moved through the heavens – in effect delivering the information that Copernicus had implied should be available for every celestial object he treated. However, Maestlin and Brahe did not thereby acquire the concept of an orbit.

The track described by Maestlin and Brahe took the comet through several of the geocentric shells or orbs accepted by Ptolemy and Aristotle, a motion that was supposed to be impossible. But neither observer concluded that there were no celestial orbs – at least not immediately. Maestlin continued to accept celestial orbs but concluded that they must be centered on the sun. He believed he had discovered that the comet itself was confined to an orb, just like one of the planets. Its orb was located outside the orb of Venus but inside that carrying the earth-moon system. The track of the comet showed only that there were no *geocentric* orbs. Brahe also assigned the comet to a heliocentric orb just larger than that of Venus, but he was unable to accept Copernicanism because of physical objections to the motion of the earth. Consequently he preferred a geoheliocentric arrangement for the planets (the earth is central, the moon and sun revolve around the earth, all the other planets and the 1577 comet revolve around the sun). In this arrangement the spherical shells for the sun and the planet Mars intersect, an outcome that is also supposed to be physically impossible if they are constituted as Aristotelians believed. In response to this problem, Tycho abandoned material shells entirely. By 1588, when he published his account of the comet, he had adopted fluid heavens in which the orbs of the planets became merely geometrical boundaries.

Neither Maestlin nor Brahe changed the concept of PATH from the established pattern we examined in Ptolemy and Copernicus. Both continued to decompose celestial motions into circles. The information they provided about distances for the comet of 1577 was used as negative evidence against Ptolemy, not as a positive contribution to astronomy. It was left to Kepler, as the intellectual heir of both men, to see the possibilities of replacing circles by ellipses, and spherical shells by orbits.

6.5 THE CONCEPTUAL STRUCTURE OF KEPLER'S ASTRONOMY

By the time Kepler began his career in the 1590s all the information we have just reviewed was readily available to him. In the *Mysterium Cosmographicum*, published in 1596, he opted for fluid heavens through which planets move freely within the confines of heliocentric shells. These shells are no longer physical but purely geometrical structures defined by his ingenious Platonic solid construction. But the motions of these planets are still decomposed into circles, and the frame for the concept of PATH that Kepler is using in 1596 would be no different from the one we considered in the case of Copernicus.

The concept of an orbit first appears in the *Astronomia Nova* of 1609. Although the book is presented as a narrative of Kepler's discoveries, and the ellipse that is the orbit of Mars does not appear until right at the end, Kepler clearly has the concept of an orbit in view from the very beginning of the book (Stevenson 1994; Donahue 1988, 1992; Voelkel 2001). On page 4 of the *Astronomia Nova* Kepler presents a picture of the orbit of Mars in Ptolemaic astronomy – that is, he draws the track of the planet as a continuous curve in two dimensions (Figure 66). Examining the changes Kepler introduced in the frame diagram for PATH corresponding to his work in the *Astronomia Nova* will show us

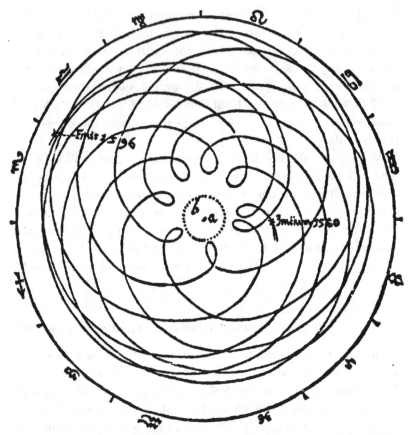

FIGURE 66. Kepler's illustration of the geocentric trajectory of Mars, 1580–1596. Ioannes Kepler, *Astronomia Nova* (1609), p. 4. Copyright the History of Science Collections, the University of Oklahoma Libraries, and reproduced by permission.

several interesting things about how concepts change and also allow us to pinpoint the first major incommensurability in astronomical concepts due to Copernicanism.

Although Kepler ultimately subverts the whole structure of the PATH frame, at the outset his changes fit entirely within the established pattern. Taking the Copernican frame (Figure 60) as a starting point, let us consider each of the three attributes and the corresponding frames in turn. Kepler's treatment of diurnal motion is the same as Copernicus', so the version of the frame for DAILY MOTION corresponding to the conceptual structure introduced in the *Astronomia Nova* will be no different.

Next, let us consider the frame for RETROGRADE MOTION. Kepler's account of retrogression is also the same as Copernicus' in general terms – although it differs in some important specifics. Copernicus had attributed a single circular motion to the earth, while all the other planets moved on eccentric circles modified by small epicycles. He thus gave a different account of the motions of the planets and the earth. Kepler insisted from the beginning of the *Astronomia Nova* that all planets including the earth should be treated in the same way. Ultimately this will mean giving each one an elliptical orbit. So the attributes and values in the frame for RETROGRESSION must differ from Copernicus', which correspond to a circular motion, and be replaced by whatever Kepler wants to say about the elliptical orbit of the earth. However, for the first forty chapters of the *Astronomia Nova* Kepler also considers only circular motions, although, ironically, he reintroduces the equant, with all of the complications it entails, before establishing a preliminary version of the second law of planetary motion applied to an eccentric circle, as the basis for calculating the values of the attribute we have called SPEED. Let us postpone a complete presentation of the frame for RETROGRADE MOTION until we have seen how the frame for the crucial attribute PROPER MOTION changes.

The most important changes in the new frame for PATH occur in the attribute PROPER MOTION. The recursive frame representing the attributes and values of PROPER MOTION was originally based on the structure of the frame for CIRCULAR MOTION (Figure 57). Modifications introduced by Kepler in his account of proper motion will affect both this recursive frame and the parallel one representing retrogression.

As soon as the material orbs of Aristotle and Ptolemy were abandoned the question of what moved the planets became acute. Kepler still accepted a version of the Aristotelian concept of inertia – objects move only while a force acts on them, and the motion is in proportion to the intensity of the force. In common with contemporaries including Bruno and Galileo, Kepler assumed that a power or force located in the physical sun was responsible for moving the planets. He believed this force diminished with distance (in our terms an inverse first power rather than an inverse-square law). As the solar force diminished with distance it would propel a planet more slowly when farther from the sun and more quickly when closer to it. If the circle on which a planet moved was eccentric to the sun, then the planet would move more slowly near aphelion and more quickly near perihelion. Kepler used this relation between distance and velocity (understood now as the velocity of the planet *along its track*) to calculate the angular position of the planet, viewed from the sun, and ultimately the earth. This set of techniques (later – and in a modified form – to be labeled Kepler's second law) replaces the attribute SPEED in the earlier CIRCULAR MOTION frame.

The initial presentation of the distance-velocity relation takes place in chapter 40 of the *Astronomia Nova* and uses an eccentric circle for the planet's track. Consequently, a frame diagram for Kepler's account of **PROPER MOTION** based on the *Astronomia Nova* up to chapter 40 would look very like the corresponding part of the frame for the structure of Copernican astronomy. The frame as a whole would have the features introduced in the frame for CIRCULAR MOTION (Figure 57). The replacement of the attribute CENTER by the attribute CENTER OF FORCE does not change the range of values attached to that attribute, although we may now select centers that correspond to particular objects, such as SUN and PLANET. The attribute RADIUS and its values might also be used without modification. The most conspicuous difference appears in the replacement of the attribute SPEED by a new attribute corresponding to the distance-velocity rule. Where before the SPEED had been a single fixed value of angular velocity for a circle or orb, the new attribute represents the instantaneous velocity of the planet along its track at a given moment in time. This is a parameter that varies continuously over time according to the rule given by the distance-velocity principle. Although a new attribute has clearly entered the frame here, all the values in the new frame are already permitted under similar attributes in the equivalent Copernican frame; hence the change is conservative. Without new attribute-value combinations there can be no incommensurability between the two frames.

But Kepler does not conclude with the circle-based exposition of chapter 40, primarily because the positions for Mars against the background of fixed stars calculated using an eccentric circle fail to agree with calculations of its distance. To accommodate the distance data Kepler introduces a second principle that governs the motion of the planet toward or away from the sun along the radial line defined by the distance-velocity relation. The planet is made to reciprocate on a diameter of a hypothetical epicycle. This is not an epicycle like those used by Ptolemy and Copernicus. The planet is depicted as moving along the diameter of the epicycle not around the circumference, and, while the motion is real, the epicycle is only a calculating device. This set of techniques allows Kepler to define the heliocentric distance to the planet at all points along its track and replaces the DISTANCE attribute in the earlier frame of CIRCULAR MOTION. And like the attribute already introduced in place of SPEED, the values of this new attribute are a set of distances that vary continuously over time as the planet moves around the sun. Thus, we can argue that the conceptual structure of Kepler's astronomy differs from that of Copernicus because in his account of PROPER MOTION he replaces all three attributes in the frame for CIRCULAR MOTION (Figure 57). CENTER has been replaced by CENTER OF FORCE. RADIUS has been replaced by a calculation of DISTANCE based on the reciprocation rule. And SPEED (an angular measure) has been replaced by a calculation of SPEED (now a linear measure) based on the distance-velocity rule. In all three cases values corresponding to the new attributes are within ranges admitted for the original attributes. But these revisions entail a further and even more important revision.

By introducing the distance-velocity rule and the reciprocation rule, Kepler is able to recover all the positional data corresponding to the concept of PATH. But he also shows (1609: Ch. 58–60) that the actual track of the planet is an ellipse inside the eccentric circle introduced earlier. Thus, the superordinate concept with attributes CENTER OF FORCE, DISTANCE, and SPEED is no longer CIRCULAR MOTION but something new. It is not merely the concept of an elliptical motion, but also the concept of a continuous, real track in space, which we may now correctly call an ORBIT (Figure 67).

As mentioned earlier, Kepler insists that the motion of the earth be treated in exactly the same way as that of other planets. But the proper motion of the earth figures centrally in the explanation of retrogression for both Copernicus and Kepler. So any changes in the

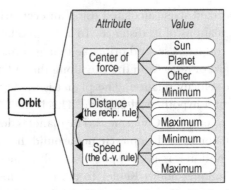

FIGURE 67. Partial frame for 'orbit' as introduced by Kepler in the *Astronomia Nova* (1609).

frame for PROPER MOTION will also appear in the frame for RET-ROGRESSION. If we now compare the frame for PATH in Copernican astronomy (Figure 60) and Kepler's version (Figure 68) we see major changes. For Kepler, the new frame for ORBIT (Figure 67) takes the place of the frame for CIRCULAR MOTION (Figure 57) when we perform the recursive expansion of the nodes for PROPER MOTION and RETROGRESSION. The resulting frames differ conspicuously from their Copernican predecessors. The Copernican and Keplerian versions of these frames are incommensurable because a whole set of attribute nodes have been replaced along with their corresponding values, creating entirely new possibilities for classifying celestial objects as planets, but also because now, for the first time, the sun plays a special role in the conceptual structure. The sun appears explicitly as one possible value for the attribute CENTER OF FORCE, but also implicitly as the center from which distance and velocity are specified in the other two nodes. The special role of the sun, and the subordinate role of the earth, become part of the overall structure.

The recursive expansion of the node for DAILY MOTION will be the same in both Copernican and Keplerian versions, because both attribute this to a rotational motion of the earth, which is a circular motion with the attributes and values depicted in Figure 57. So the introduction of the concept of an orbit by Kepler creates a division between the concept of daily motion and the concepts of proper motion and retrogression. A further division is introduced by considering whether these motions are real or apparent. Over the course of

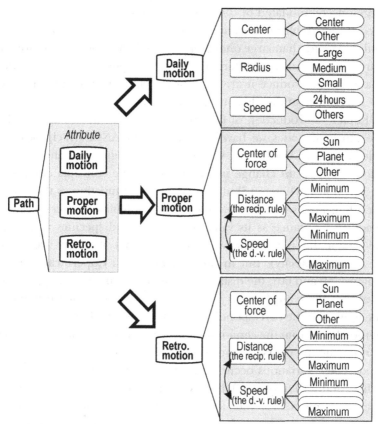

FIGURE 68. Partial recursive frame for 'path' showing modifications introduced by Kepler in the *Astronomia Nova* (1609).

subsequent centuries these divisions became the basis for a complete separation of the concept of daily motion and retrograde motion from the concept of path. The modern concept of 'path of a planet' is synonymous with 'orbit of a planet'.

Kepler later dispensed with the separate agencies that underlie the distance-velocity rule and the reciprocation rule. The *Epitome of Copernican Astronomy* (1618–1622) derives both the transverse and the radial motion of a planet from a single force originating in the sun. But it is probably significant that his original argument distinguishes between these features of the conceptual structure, and that the original presentation of the distance-velocity rule used a circle and could be largely accommodated by the existing conceptual structure, without incommensurable modifications. Small changes are easier to conceive, and to defend, than large ones. Kepler's final conceptual structure for astronomy is radically different from the Copernican one, but the changes that produce it are incremental and initially fit within the overall pattern of the old conceptual system.

Kepler's introduction of the concept of an orbit, and specifically the attribute nodes we have designated DISTANCE (reciprocation rule) and SPEED (distance-velocity rule) in Figures 67 and 68, signals an important change in the nature of the concepts employed in astronomy. These attributes serve the same function as the corresponding nodes in the Copernican and Ptolemaic frames: the first specifies a distance as its value and the second specifies a velocity (although this is now the velocity of a planet along its orbit, not its angular velocity). But in the Ptolemaic and Copernican frames (Figures 59 and 60), the distance, or angular velocity, that appears as a value of these attributes is *fixed* and takes a single value throughout the motion of the corresponding planet. For Kepler both the distance of the planet from the center of motion and its velocity along its orbit *vary* during the course of each complete orbit and repeat during the next. These variations occur within fixed ranges (defining different orbits for different planets), but a specific, single value of distance, for example, now corresponds to a value of velocity that may exist only for an instant (or, given that the orbit is symmetrical, two instants, one approaching the sun, the other receding from it). The frames we have drawn for object concepts allow only a single fixed value for each attribute. This kind of diagram works perfectly well for Ptolemaic astronomy or the original version of Copernican astronomy. But if we used frames that display only a single activated value for each attribute for the case of Kepler, the concept of an orbit would have to be represented by a series of frames, each with a specific value for distance and velocity at different times. At different moments in the planet's motion the frames in the series would succeed one another according to a fixed pattern (corresponding to the ranges of linked values allowed for distance and velocity in the orbit of that planet). Concepts represented by a series of frames form an important element in scientific conceptual structures; however, they require separate consideration.

Cognitive scientists distinguish broadly between *object concepts,* with values that do not vary over time, and *event concepts,* which may embody values that vary over time and may require multiple frames in their representations. The concept ORBIT is the first event concept we have encountered in this study; all previous concepts were object concepts that could be adequately represented by a single frame with time-independent values. Experiments by Barsalou and Sewell (1985) suggest that when event concepts are represented, memorized, and retrieved, they are processed in a way that differs from object concepts. Specifically, it seems that the temporal relations inherent in event concepts are not represented by properties, but by dimensional organizations of temporal sequence, or chronological orders, like the time-ordered sequence of positions and velocities that constitutes an orbit. If this is the case, any attempt to represent chronological orders by properties (attributes) alone is not only inefficient but inaccurate.

Currently, cognitive scientists still understand little about how chronological orders are conceptualized. According to Barsalou, one possibility is that some kind of framelike structure is used to produce the successive states covered by an event concept. Some cognitive scientists call this kind of complex framelike structure a mental model, but they have many different interpretations of the nature and character of this cognitive structure. Barsalou uses a modified frame structure (Figure 69) to represent the event concept ENGINE CYCLE (Barsalou 1992b: 55). In general, crossing a frame for an object and a frame for time provides a means of representing event concepts. On the left-hand side of Figure 69 is a *component frame* and, as in the case of frames for object concepts, its attributes are the major parts of an engine (IGNITION, INTAKE VALVE, EXHAUST VALVE, and PISTON). Unlike the frames for object concepts, however, the values in this frame represent different states of operation. For example, CHARGING and SPARKING are the two states of IGNITION, and COMPRESSING and DECOMPRESSING are the two states of PISTON. The component frame also indicates that there are constraint relations among attributes – INTAKE VALVE, EXHAUST VALVE, and PISTON are in fact mechanically connected.

On top of the right-hand side of the figure is a *sequence frame* that captures both the temporal order and the causal connections of the

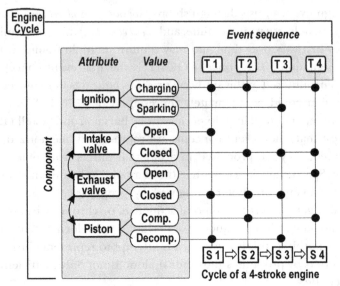

FIGURE 69. Partial frame for the event concept 'engine cycle'. Redrawn from Barsalou 1992b.

event sequence. The four attributes of this frame represent four different moments in the sequence, each of which takes a specific value corresponding to an attribute in the component frame. For example, T1 in the event sequence takes four specific values from the four attributes in the component frame: CHARGING IGNITION, OPEN INTAKE VALVE, CLOSE EXHAUST VALVE, and DECOMPRESSING PISTON. Thus, by crossing two frames and noting all the intersections, we obtain a sequence of subordinate concepts STROKE 1, STROKE 2, and so on, which collectively represent a specific event – THE CYCLE OF A FOUR-STROKE ENGINE.

A similar structure could be used to represent the concept ORBIT for Kepler and his successors, by means of a sequence of frames each of which corresponds to a unique combination of distance and direction of a planet from the center of force. In Figure 70 the attributes in the component frame, on the left, are the same as the attributes introduced in Figure 67. The double-headed arrow indicates that the values of the attributes distance and speed are linked (by the distance-velocity rule, or Kepler's second law). Although there are in

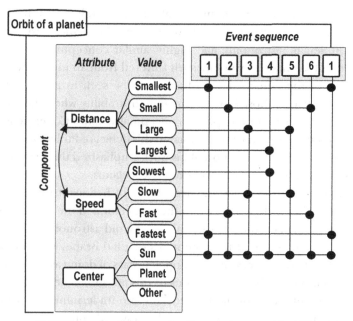

FIGURE 70. Partial frame for the event concept 'orbit'.

principle an indefinitely large number of these linked values, for purposes of illustration we consider only four, corresponding to the maximum and minimum values of each, and two intermediate values. The sequence frame on the right shows seven moments in the pattern corresponding to a single orbit: as the planet moves away from closest approach to the sun (DISTANCE: SMALLEST) its speed diminishes to a minimum at its largest distance (SPEED: SLOWEST) and grows again to a maximum at its smallest distance; then the entire cycle repeats.

6.6 INCOMMENSURABILITY, INCREMENTAL CHANGE, AND THE COPERNICAN REVOLUTION

In this chapter we have examined the conceptual structures of Ptolemaic astronomy, Copernican astronomy in its original form, and Kepler's variant as it appeared in the *Astronomia Nova* of 1609. This has led to a dramatic reappraisal of the Copernican revolution.

Rather than a clear incommensurability with Ptolemaic astronomy introduced by Copernicus, we have seen that Copernican astronomy and Ptolemaic astronomy had highly similar conceptual structures, and that Ptolemaic astronomy itself suffered from an internal defect, the use of the equant device, that could be seen as a difficulty of the same type as those labeled incommensurability when they occur between successive or competing conceptual systems. Copernicus was widely credited with having eliminated this defect of Ptolemaic astronomy, and the initial reception of his work emphasized this feature and disregarded his more radical cosmological claims.

We have suggested that the acceptance of heliocentrism as a cosmological theory depended upon the elimination of the celestial orbs that were the main theoretical entity of the old astronomy. Although we have traced some of the arguments that led Brahe, Maestlin, and Kepler to abandon orbs and develop new ways of doing astronomy, we should emphasize the partial status of this account. The Aristotelian-Ptolemaic substance of the heavens was also undermined by parallel developments in fields as diverse as alchemy and optics.

Kepler's work was the first sustained defense of Copernicanism in the modern sense: the sun played a real physical and geometrical role, and planets moved around it on paths that could be calculated from Kepler's new principles. Their speed, distance, and direction could all be known. Orbs had been replaced by orbits (Barker 2002). Comparing the conceptual structures used by Kepler and those of his predecessors, we conclude that the first major incommensurability with earlier astronomy occurred in the work of Kepler, but he breaks equally with both Ptolemy and Copernicus. Kepler's concept ORBIT (Figure 67) is incommensurable with the earlier concept CIRCULAR MOTION (Figure 57) that is used in the recursive expansion of PATH for a celestial object (Figure 58). We have also suggested that to be adequately represented Kepler's concept should not appear as a single frame with fixed values but rather as an array of frames showing how linked values vary over time. While PATH was an object concept, ORBIT is evidently an event concept, and the displacement of the former by the latter may mark a discontinuity between ancient and modern astronomy at least as important as the incommensurability between the PATH and ORBIT.

Last, these changes did not occur at a stroke. The changes in the conceptual structure of astronomy occurred incrementally. These changes also led to the development of new conceptual resources, such as the concepts of an orbit and of a center of force, that were incorporated piecemeal into the wider conceptual structure for CELESTIAL OBJECT with which our discussion began.

7

Realism, History, and Cognitive Studies of Science

In this final chapter we attempt to do three things. First, we review the results we have presented. Next we consider the implications of our position for one of the major controversies within philosophy and sociology of science, the realism debate. Finally, we consider the significance of our results for wider debates in the history, philosophy, and sociology of science.

7.1 RESULTS

Our goal throughout this book has been to recover and extend Kuhn's account of scientific change by showing that its most important features are consequences of the nature of concepts, as currently understood in cognitive psychology and cognitive science. An important subsidiary point is that Kuhn's own theory of concepts has been shown to be independently supported by work in cognitive psychology and cognitive science.

We have shown that there is a defensible distinction between normal science and revolutionary science, but that the difference between them is not a question of the historical rarity of one process versus the other. Viewed as conceptual changes both processes may occur at any time. In the case of revolutionary change, whether the result is a big revolution or small one depends on other factors – the status of the conceptual structure that changes (for example, whether or not the changes affect a fundamental item in the ontology of the field), as well as the speed and completeness with which the changes are adopted. All these are factors that may be influenced by other forms of historical causation in addition to purely cognitive issues.

Kuhn's original mechanism for initiating change, the anomaly, is vindicated by our account. A detailed understanding of the mechanisms by which anomalies bring about conceptual change is available by considering the violation of the hierarchical principles we have described, and especially the no-overlap principle introduced by Kuhn himself.

In the case of revolutionary changes, we have shown, first, that the entire conceptual structure does not have to change when they occur; second, that they do not happen instantaneously; and, third, that they do not guarantee communication failure. Examples from the development of ornithology during the Darwinian revolution show all these features, and that common features of incommensurable conceptual structures may still furnish a basis for rational comparison.

Our account shows clearly and in detail the origins and nature of communication failure between scientific communities, to the extent that it is based on discrepancies in conceptual structure. The discovery of nuclear fission required a revolutionary modification in an existing conceptual structure for classifying induced nuclear decays. The fate of Ida Noddack's 1934 suggestion, which might today be counted as the proposal of a fission process, is better understood, from the viewpoint of those who were committed to the conceptual structure that was actually modified to accommodate the discovery of fission, as an unwarranted proposal of a revolutionary change in that structure. A similar point may be made about the Copernican revolution. We have shown that, from the viewpoint of astronomy considered as a separate science, the conceptual structure used by Copernicus is continuous with that employed by Ptolemaic astronomers, and this continuity corresponds to the way in which it was actually received for several decades after Copernicus' death. Although the cosmology proposed in Copernicus' work is revolutionary, the changes in existing conceptual structures they require may also be regarded as unmotivated, or unwarranted at the time that Copernicus introduced them. Adopting a Copernican cosmology required the unseating of an established ontology of celestial substances. The actual evidence competent to question and remove the ontology of celestial orbs did not appear until after Copernicus' death, and although some of it was gathered by astronomers, a number of other fields have reasonable claims to stakes in this negative argument.

Rather than regarding the response to these proposals – Noddack's premature proposal that the atomic nucleus could divide into two, and Copernicus' cosmology – as instances of communication failure, it is perhaps more useful to see them as instances of resistance to changes in conceptual structures. Whatever other factors are at work in motivating this resistance, our analysis shows a definite cognitive element in such situations. In Noddack's case the conceptual divergence between her proposals and the accepted conceptual structure made it difficult or impossible for contemporaries to take her proposal seriously. Copernicus benefited from eliminating a difficulty of this kind in Ptolemaic astronomy (the equant problem), but the astronomical theory that led directly to the modern version of Copernicanism, developed by Kepler, was provided in the form of an incommensurable cosmology supporting an incommensurable astronomical theory. A further liability was the replacement of the existing concept of the path of a planet, which had been based on the concept of circular motion and had the structure of an object concept, with a new account based on the concept of an orbit, an event concept. Kepler's version of Copernicanism suffered resistance on a scale even greater than Noddack's proposal, and was only vindicated by the general acceptance of Newton's theories more than a century later.

Throughout this study we have built on ideas provided by psychologists, and especially cognitive psychologists, to examine issues of interest to historians, sociologists, and philosophers of science. The continuing use of concepts like incommensurability among psychologists, historians, and sociologists contrasts with the continued denial by many philosophers of science that incommensurability is a real phenomenon or that the preferred account of science should accommodate it. Part of the problem has been that many earlier studies treated incommensurability as abnormal or as an obstacle to progress or an obstacle to the rational evaluation of scientific theories. We have suggested solutions to all these problems.

In our account incommensurability may be seen as a natural accompaniment of certain kinds of change in any conceptual system – it is not limited to science. The techniques we offer also provide a means for locating and appraising the degree of incommensurability.

Incommensurability will appear where changes in attribute-value sets create new similarity and difference relations among objects; existing objects are reclassified while new objects excluded from earlier conceptual systems may appear. The degree of incommensurability will depend on how extensive this revision is, how high up in a conceptual hierarchy the revised concepts appear, and whether these revisions require the modification or elimination of constraint relations. The most important of these, in turn, will be the constraints that represent causal theories. We therefore suggest that the further study of the role of causal theories as constraints in conceptual structures is an important area for additional research.

The changes that generate incommensurability may be local rather than global as they apply to both conceptual structures and the corresponding scientific communities, but incommensurability itself is a consequence of the nature of conceptual structures and hence a universal feature of the human use of concepts. Incommensurability can arise when any conceptual structure changes. The examples from science are unusually striking not because the phenomenon is unique to scientific change but more because science is committed to public discussion of revisions to conceptual structures in a way that other human practices are not.

The replacement of object concepts by event concepts suggests a range of new problems that may also generate incommensurability. However, the study of these two categories of concept is only at a preliminary stage, and more work is clearly required. In ordinary life, the two types of concept interpenetrate. Although we have largely treated 'bird' as an object concept, it contains attributes ('gait', for example) that are themselves event concepts. And just as engine cycles contain a number of stages corresponding to piston strokes, all biological organisms might be regarded as sequences of events with important stages like birth, growth, maturity, reproduction, death, and in the case of birds an annual cycle of plumage changes. So one and the same concept may appear an object concept for some purposes but an event concept for others.

The transition from astronomy as developed by Ptolemy and Copernicus to astronomy as developed by Kepler appears to be a relatively clear case of the replacement of an object concept by an event concept doing the same work. Other examples include the replacement of the concept of polarization in the particle theory of light by the concept of polarization in the wave theory of light, during the early nineteenth century (Chen 2003). But it would be premature to conclude that such transitions always occur in one direction. There may equally be examples of historical transitions from event concepts to object concepts (the change from James Clerk Maxwell's theory of charge to Hendrik Antoon Lorentz' theory of the electron may be an example). The questions that emerge from our study are when, where, and how such transitions take place.

A further important application of cognitive theories of concepts may be in understanding how the goals of scientific research influence conceptual structure. Goals and interests are invoked by sociologists of science to explain how scientists choose between different courses of action and different concepts (Barnes 1982: 102–103), but little attention has been paid to the cognitive status of goals. Work in cognitive science has produced several different models of the way goals influence conceptual structures (Barsalou 1982; Smith 1988). This work indicates that goals and interests are not the overriding explanatory factors suggested in some sociological accounts of science (see Section 7.3) but contribute along with other cultural and especially cognitive factors to determine historical outcomes. Thus, this research supports our general position that explanation of change in science requires a mix of social or historical factors with cognitive factors. Regardless of the direction these and the other research questions we have raised take in the future, we are confident that the understanding of science requires renewed attention to its cognitive structure.

7.2 REALISM

Our account of incommensurability and the related claim that entities may appear or disappear during the development of science raises questions regarding the ontological status of scientific objects. In this section we shall first review the realist response to the incommensurability thesis. We shall then explain why we reject the realist view and instead see scientific concepts as referring to entities in a phenomenal world. We shall describe how, on this view, anomalies may trigger referential changes, and we shall argue that on this view chain-of-reasoning arguments secure the comparability between incommensurable theories.

7.2.1. Incommensurability and Realism

As we have discussed in Chapter 5, the incommensurability thesis asserts that successive theories employ different conceptual systems and that, consequently, some of the terms that may seem to be shared by the competing theories may differ in meaning. By the same token, competing theories may posit entities that do not exist according to their competitors. Because of such differences in meaning and in ontology some – but not all – statements of the one theory cannot be translated into statements of the other without residue or loss. Therefore, to the extent that claims of the one theory cannot be translated into claims of the other, the content of these claims cannot be directly compared.

An important response to this version of the problem of incommensurability has been the referential stability approach. This approach was first proposed by Scheffler (1967), who argued that even if two theories are mutually untranslatable, as long as their terms share reference it is possible for statements from the two theories to conflict, and hence for the theories to be rivals and comparable. Several scholars, most notably Putnam, have therefore argued for a theory of reference that would enable a term to refer although scientists' beliefs about it might be mistaken. Putnam argued that the causal theory of reference would achieve this kind of referential stability. According to the basic version of the causal theory, a term is introduced in an original naming ceremony. At this introduction, the object or kind to which the term shall refer is singled out by ostension or by a description. In subsequent use the term continues to refer to the entity to which it was originally attached on the occasion of its introduction. For kind terms, the extension of the term is fixed by means of a representative sample, and the extension of the term consists of the set of objects that bear the same-kind-as relation to objects in the original sample. The same-kind-as relation is taken to be a theoretical relation determined by the internal structural traits of the objects to which the term refers, and the details of the relation are thus to be discovered by scientific research (Putnam 1975: 225; Boyd 1979; Sankey 1994: 52).

The referential approach makes several realist assumptions: it presupposes that some fixed realm of "theory-independent entities" exists

(Putnam1975: 236) and that the aim of science is to improve the accordance between our concepts and these entities, to "cut the world at its joints" (Boyd 1979: 483, similarly Putnam 1975). On this view terms are used "as if the associated criteria were not *necessary and sufficient conditions,* but rather *approximately* correct characterizations of some world of theory-independent entities"; hence, later theories are "in general, *better* descriptions of the *same* entities that earlier theories referred to" (Putnam 1975: 237, italics in the original). However, as we shall show in the next section, this is exactly the kind of realism called in question by the incommensurability thesis.

7.2.2. Entities in a Phenomenal World

As we have described in Chapter 2, on Kuhn's view – and ours – the division of the world into kinds is constituted by a web of similarity and dissimilarity relations. Kuhn claimed that these relations are "primitive" (Kuhn 1970a: 200) or "immediate" (Kuhn 1970a: 197, fn. 14) in the sense that they are not based on a relation that confers similarity to the entities it links. Further, the immediacy of the similarity and dissimilarity relations is made possible because of an "empty perceptual space between the families to be discriminated" (Kuhn 1970a: 197, fn. 14; see also Section 2.3 and Ch. 4). At first sight this may seem to resemble a realist position in which the subdivision of the world can simply be read off the world itself. However, this is not the case. Perceptual space is spanned by a multitude of features and there need not be one unique division of this space. In pre-Copernican astronomy, the moon and sun, as well as Mercury, Venus, Mars, Jupiter, and Saturn, were classified as 'wandering stars' or 'planets' because they moved against the fixed pattern of background stars. In post-Copernican astronomy, and especially in astronomy after Kepler and Newton, these objects are divided into three classes (stars, planets, and moons), this time on the basis of the kind of orbit they have. To emphasize that there is not one unique subdivision of the world into entities, but that the world may be perceived as consisting of different entities dependent upon which features are considered important, we shall call the perceptually and conceptually subdivided world a *phenomenal world* (Hoyningen-Huene 1993).

A similar position has been taken by cognitive psychologists like Rosch and her collaborators, whose research has been based on the working assumption "that (1) in the perceived world, information-rich bundles of perceptual and functional attributes [which here means features used by subjects performing categorization tasks] occur that form natural discontinuities, and that (2) basic cuts in categorization are made at these discontinuities" (Rosch 1978: 31). Although Rosch admits that she originally took the realist view that the features inhered in the real world (Rosch 1978: 41; Rosch et al. 1976; Rosch and Mervis 1975), she later realized that some types of features have meaning only in relation to other categories (for example, attributes such as 'small' or 'large'), and other features require knowledge about the object as an instance of the concept in question, or knowledge about humans and their activities in order to be understood (for example, features such as 'seat' or 'handle', which require knowledge of the use of the objects in question). Hence, for such features "it appeared that the analysis of objects into features was a rather sophisticated activity that our subjects (and indeed a system of cultural knowledge) might well be considered to be able to impose only *after* the development of the category system" (Rosch 1978: 42). To put it simply, the discontinuities between bundles of attributes determine conceptual structure.

For Kuhn's and Rosch's positions the mutual dependence between the entities in a given phenomenal world and the relations of similarity and dissimilarity may seem circular: the entities in the form of bundles of attributes secure the immediacy of the relations of similarity and dissimilarity, but the relations of similarity and dissimilarity are constitutive of the entities. However, the circle can be made to vanish by adopting a developmental or historical view. On such a view, the world has not been structured from scratch by its inhabitants. Instead, the entities of a phenomenal world are inherited by any generation from their predecessors and are therefore in place, ready to secure the immediacy of the relations of similarity and dissimilarity for the new generation. But once the new generation has gained access to this world they may start reshaping it by introducing new relations of similarity and dissimilarity and abandoning old ones and thus leave to their successors a different set of entities from the set they inherited themselves.

7.2.3. Anomalies and Restructuring of the Phenomenal World

According to the position developed here, a world with its entities is constituted by relations of similarity and dissimilarity between objects. However, although different sets of similarity and dissimilarity relations may constitute different worlds, they cannot be freely invented to constitute any arbitrary world.

Both Rosch and interpreters of Kuhn have argued that occasionally situations occur in which it becomes clear that something is wrong with the structure that our concepts give to the world – that objects do not behave or situations do not develop as prescribed by the current conceptual structure. Thus, Rosch argues that the structure of the world is constrained by certain real-world factors: "In the evolution of the meaning of terms in languages, probably both the constraint of real-world factors and the construction and reconstruction of attributes are continually present. Thus, given a particular category system, attributes are defined such as to make the system appear as logical and economical as possible" (Rosch 1978: 42). These real-world factors function as constraints in the sense that they offer resistance against giving arbitrary structures to the world: "If such a system becomes markedly out of phase with real-world constraints, it will probably tend to evolve to be more in line with those constraints – with redefinition of attributes ensuing if necessary" (Rosch 1978: 42).

Likewise, Hoyningen-Huene in his reconstruction of Kuhn's position argues that for anomalies to occur, "the phenomenal world under study must also exhibit a certain independence from theoretical expectations, otherwise such discrepancies between theory and experience couldn't exist" (Hoyningen-Huene 1993: 226–227). This 'certain independence' Hoyningen-Huene ascribes to a *world-in-itself* that, like Rosch's real-world factors, functions as a constraint against giving arbitrary structures to the world: "The resistance of the world-in-itself (or of stimuli) may, to some extent, penetrate the network of similarity relations" (Hoyningen-Huene 1993: 227).

We shall follow Rosch and Hoyningen-Huene in arguing that anomalies may reveal that something is wrong with the structure that our concepts give to the world. As we argued in Section 4.3 this may happen when an object is encountered that violates the hierarchical principles: that is, when judged from different features it will be categorized into different contrasting categories. When this happens we see that the expectations about how objects behave or how situations develop, which are conditioned by our existing conceptual structure, do not hold, and that we must change our conceptual structure in response.

For example, seventeenth-century ornithologists identified water birds by their round beaks or their webbed feet (as described in Chapter 4). The equivalence of the two features suggests an empirical correlation between them. The conjunction of these features can therefore be seen as a hypothesis about the behavior of instances of the corresponding concept – in this example, the hypothesis that all birds with webbed feet also have rounded beaks. But whether this hypothesis holds is an objective matter. Anomalies will appear if the features are not correlated after all: that is, instances will be discovered that show that the correlation does *not* hold. For example, the screamer with its webbed feet and pointed beak (Figure 19) showed that the hypothesis does not hold. However, it is important to note that the preceding objectivity claim only implies that one cannot make arbitrary hypotheses about the behavior of instances of a given concept. It does not rule out alternative hypotheses. As explained in Chapter 4 the Sundevall and Gadow classifications of birds imply different hypotheses about which categories of bird exist and how these categories are characterized. Thus, as argued previously, different sets of similarity and dissimilarity relations may constitute different ontologies, that is, posit different entities in the phenomenal world. Hence, the objectivity claim does not assert the existence of entities in what philosophers have called the *real* world, or the world-in-itself.

7.2.4. Chain-of-Reasoning Arguments, Conceptual Continuity, and Incommensurability

On our developmental view, the structuring of the world is the result of a historical process that incorporates changes to previous structurings of the world in response to anomalies. However, as explained in the previous section, an anomaly reveals only that specific features are not correlated; it does not determine which features are correlated instead. The only requirement is that the new combination of features can be seen as hypotheses with some positive, but no negative cases.

Anomalies and the requirement of establishing new hypotheses on the correlation of features, which may have some positive but no negative cases, thus provide *reasons* for creating new hypotheses on the correlation of features.

We follow Shapere (1989) in arguing that continuity of reference can be established by there being reasons for changing the features ascribed to an entity. On Shapere's view, if "there is a chain of reasoning in terms of which we can understand why certain properties ascribed in usage U_1 and its successors up to and including U_2 were abandoned, altered, or replaced, then that chain-of-reasoning connection explains the possibility of comparing the two usages and their theoretical contexts despite the fact that, within each, the usages of the term or terms involve few or even no properties in common" (Shapere 1989: 417). Thus, the different historical stages in the development of a concept may also be seen as a family resemblance class with a complicated network of overlapping and crisscrossing relations between the historical stages in the concept's development. An example is the abandonment of the Averroist insistence that all celestial motions were centered on the earth. Galileo's discovery of the moons of Jupiter provided a chain-of-reasoning argument to allow other centers of motion, which proved decisive where arguments by earlier Ptolemaic astronomers had not. These diverse centers of celestial motion were carried forward historically, not in their original guise as the centers of circular motions, but ultimately as centers of force for orbital motions. Hence, on our view, continuity and comparability follow from the gradual development in which anomalies provide reasons for abandoning some features and adopting others in the conceptual structure.

7.3 THE SYMMETRY THESIS

We began this study by noting not only the general neglect of Kuhn's work, but also the sad state of cognitive analysis in science studies generally. Our epigraph from Bruno Latour and Steve Woolgar's postscript to the second edition of *Laboratory Life* may originally have been offered more humorously than seriously. But it remains the case that many people in science studies, and particularly those influenced by the more recent movements in sociology of science such as the Strong Programme and Latour's actor-network theory, believe that a more or less complete account of the history of science can be constructed without considering what we have called cognitive factors. It is therefore important for us to attempt to be as clear as possible about the status of our own proposals within the various methodologies currently available in science studies.

The contemporary devaluation of cognitive studies of science may be dated approximately from the appearance of David Bloor's manifesto for the Strong Programme in the sociology of knowledge (Bloor 1976/1991). Bloor was favorably disposed toward Kuhn, as he was then understood. But, here and in later work, Bloor criticized a succession of historically oriented philosophers of science, starting with Imré Lakatos and, after his death, turning to Larry Laudan. It is important to separate two strands in this criticism. First, Bloor may be seen as criticizing the work of philosophers as bad history. This makes it an interesting question to what extent Bloor's own form of explanation and similar work in the sociology of scientific knowledge satisfy the standards that would be expected in writing history. Second, he attacks philosophers' excessive reliance on a version of cognitive analysis and attempts to replace it with a particular form of sociological explanation, which later came to be known as Interest Theory. We believe that the kind of cognitive analysis developed in this book avoids Bloor's criticisms of earlier cognitive accounts and satisfies Bloor's requirements for historical explanation at least as well as his own proposals. We also believe that without denying a role for social causes in understanding scientific change, our analysis shows that a range of cognitive factors are needed to understand the most important kinds of scientific change, specifically those that we have called revolutionary change, and those that introduce incommensurability.

Let us consider Bloor's proposals, both from the viewpoint of general historical method and from the viewpoint of the kind of analysis we have proposed and developed in this book. In a famous passage in the opening section of *Knowledge and Social Imagery* (2nd ed. 1991: 7) Bloor proposes the following four standards for the preferred account of science:

1. It would be **causal**, that is, concerned with the conditions which bring about belief or states of knowledge. Naturally there will be other types of causes apart from social ones which will cooperate in bringing about belief.

2. It would be **impartial** with respect to truth and falsity, rationality or irrationality, success or failure. Both sides of these
   dichotomies will require explanation.
3. It would be **symmetrical** in its style of explanation. The same
   types of cause would explain say, true and false beliefs. And,
4. It would be **reflexive**. In principle its patterns of explanation
   would have to be applicable to ... itself. Like the requirement
   of symmetry this is a response to the need to seek for general
   explanations.

Let us consider these requirements in the reverse of the stated order,
considering first the extent to which each principle conforms to
generally accepted standards for conducting historical research and,
second, the extent to which frame analysis meets the same or similar
standards.

The requirement that the preferred account of science be reflexive
is, according to Bloor, a condition required by "the need to seek for
general explanations," but it is more obviously a way of deflecting a
likely philosophical objection, that sociologists of science might be
advocating standards for explaining historical events that are violated
by their own account. So the requirement of reflexivity is satisfied if
the beliefs, theoretical commitments, and so forth, of the proponents
of the Strong Programme can be explained as social constructions by
Interest Theory. Whatever its status, the requirement of reflexivity is
uncontroversially accepted as a standard in general history. The history
of history is an important study in its own right, and most historians
would accept that whatever account they give of historical causation
applies equally to the production of their own work. Similarly, the
account of concepts and conceptual change that we have proposed
may be applied reflexively: it is quite permissible to draw the frame
of the concept 'frame', and this may be useful as a way of comparing
frame theory with other accounts of concepts.

The requirement that an account of science be symmetrical in its
style of explanation is considerably more interesting, although it may
be redundant from a logical viewpoint, if the two other requirements of
causality and impartiality are accepted. Bloor glosses this requirement
as "The same types of cause would explain say, true and false beliefs."
This requirement counts against philosophical reconstructions of

history, like those of Lakatos and Laudan, that invoke one pattern of explanation (nonempirical, rational argument) to explain historical events that conform to a prescribed standard of rationality and another pattern (empirical causes, including psychological and sociological causes) to explain everything else. The former usually correspond to scientific beliefs held true today and their obvious historical antecedents ("The earth is a planet"). The latter usually correspond to beliefs held false today and perhaps not deemed scientific ("The position of the planets when someone is born influences that person's character"). Adoption of the symmetry condition places adherents of, for example, Interest Theory in a position to explain the historical development of many subjects that were once unquestionably deemed scientific but that do not exist today.

In general historical work, true and false beliefs are regarded as equally amenable to historical explanation. The rejection of the symmetry principle in some historically oriented philosophy of science is a major divergence from accepted standards of good historical explanation. However, the account we have proposed is clearly symmetrical: truth and falsity of beliefs are irrelevant to the conceptual structures presented in frame analysis. Most or all of the beliefs about astronomy that we have attributed to followers of Averroes, Ptolemy, and Copernicus (before the work of Kepler) are false.

The symmetry condition is largely an application of the more sweeping requirement of impartiality, that the preferred account of science deal in the same way with both sides of the dichotomies: truth or falsity, rationality or irrationality, and success or failure. The application of this condition led adherents of the sociology of scientific knowledge to produce notable studies of defeated theories in the history of science, for example, phrenology (Shapin 1975) and Hobbes' physics (Shapin and Schaffer 1984), but their work has applied less conspicuously to victorious ones (except perhaps Pickering (1984) on quarks and Rudwick (1985) on geology). Again, this standard is uncontroversially accepted in general historical work, and its rejection in philosophy of science is a significant lapse from standards of good historical method. However, the theory of frames is equally applicable to scientific successes and failures. Most cases of scientific change require the examination of both. As we have presented the historical development of ornithological classification in Darwinism, the discovery of nuclear fission, and the conceptual changes in astronomy during the Copernican revolution, these cases need to be understood by examining the conceptual structure that is ultimately rejected, as this forms the basis for the modifications that are incorporated into what later becomes a historical success.

The final requirement, that the preferred account of science be causal, initially separated the sociology of science most sharply from aprioristic 'logic-based' philosophical accounts. Although superficially identical to a common requirement of historical method (see, for example, Carr 1961; Evans 1999), ironically, the application of this requirement by sociologists of science has done more than any other to distance their work from standard history.

All historical explanation is multicausal, although the concepts of cause invoked in historical explanations may not be those expressed in the exact sciences through mathematical notation or in the social sciences through statistical analysis. Universal causes, and hence universal laws, are generally absent. Context plays a primary role in explanations. By contrast, the sociology of science, and especially the Strong Programme, is only doubtfully causal, except to the extent that it gives historical explanations of the type found in general historical writing. Moreover, ahistorical concepts of 'interest' and 'social construction' generally hold explanatory priority over context. And the priority given to explanations in terms of interests in the Strong Programme implies the existence of universal causes or laws of "human interest" that are alien to historical method (Barker 1998). It would be a different matter if these universal causes or laws of "human interest" were empirically well founded, but the main theoretical defense given for them, again by Bloor (1983, 2002), is a philosophical theory based on his reading of Wittgenstein, and this is separately objectionable on other grounds (Bourdieu 2001: 158–160).

The account of concepts and conceptual systems in terms of frames is uncontroversially causal to the extent that it is based on empirical research, much of which we have described. Because our account is based on empirical sources, we cannot claim that it is final. Empirical studies could undermine the models we have been using, just as they have established them. In contrast to the sociology of knowledge, we can offer good empirical reasons for generalizing from modern studies across cultures to historical cases. Frames, and the conceptual structures they represent, appear to be human universals. But frames define the structure of conceptual systems, not the content; hence context retains a primary role in historical explanations. It is an examination of the historical context that has enabled us to draw frames showing the meaning of 'path' as it applied to celestial objects before Kepler introduced the concept of an orbit, or of 'induced nuclear reaction' before the discovery of fission. Frames provide an empirically validated tool for understanding the conceptual systems of historical actors and communities, without simply imposing twenty-first-century norms, or ahistorical norms, on the investigation.

We conclude that the conditions introduced by Bloor, and generally adopted by sociologists of scientific knowledge, may be read most charitably as attempts to inject historical standards into the discussion of the history of science by philosophers. But the Strong Programme does not live up to its own standards particularly well in the central area of historical method: historical causation. By contrast, frame theory does at least as well as the Strong Programme according to these standards, and has the additional advantages that it permits the discussion of the content of science while retaining a primary historical role for context.

These considerations make our cognitive account a viable competitor to the sociology of knowledge, for example, in the form of the Strong Programme, but do not yet establish that something is missing from all accounts of the sociology of knowledge, and that what is missing is cognitive. This final point is best approached by recapitulating some of the results we have presented in the course of the book. Although the detailed account we have given of the conceptual changes that occur during normal science might in itself be offered as an argument for preferring a cognitive mode of analysis, a suitably unrepentant sociologist might insist on reconstructing the whole story in terms of social causes. However, the phenomenon of incommensurability presents a different case. Here there is an ineliminable cognitive aspect to the confrontation of different viewpoints and different groups. This emerges most clearly in the cases, like the Noddack episode, in which it is quite simply impossible for those using a particular cognitive structure to understand suggestions that violate the structure in ways we have described. During normal science a good deal of effort goes into justifying the structure of the current conceptual scheme, including modifications to it. Even if these activities are explained primarily in social terms, the activities of scientists confronted with a revolutionary change in conceptual structure involve more than social construction. Scientists cannot socially construct what they do not understand, and to explain what they do and do not understand, and how a radical change in scientific understanding comes about, requires that we admit a cognitive component in historical explanations.

References

Ahn, W. K. 1998. Why are different features central for natural kinds and artifacts? The role of causal status in determining feature centrality. *Cognition* 69: 135–178.

Ahn, W. K., & Dennis, M. J. 2001. Dissociation between categorization and similarity judgement: Differential effect of causal status on feature weights. In U. Hahn & M. Ramscar (eds.): *Similarity and Categorization*, pp. 87–107. Oxford: Oxford University Press.

Ahn, W. K., Kalish, C. W., Medin, D. L., & Gelman, S. A. 1995. The role of covariation versus mechanism information in causal attribution. *Cognition* 54: 299–352.

Amaldi, E. 1984. From the discovery of the neutron to the discovery of nuclear fission. *Physics Reports* 111: 1–332.

Andersen, H. 1996. Categorization, anomalies and the discovery of nuclear fission. *Studies in History and Philosophy of Science* 27: 463–492.

Andersen, H., Barker, P., & Chen, X. 1996. Kuhn's mature philosophy of science and cognitive psychology. *Philosophical Psychology* 9: 347–363.

Apian, P. 1540. *Petri Apiani Cosmographia*. G. Phrysius (ed.). Antwerp: A. Berckman.

Ariew, R. 1987. The phases of Venus before 1610. *Studies in History and Philosophy of Science* 15: 213–226.

Ariew, R. 1999. *Descartes and the Last Scholastics*. Ithaca, N.Y.: Cornell University Press.

Armstrong, S., Gleitman, L., & Gleitman, H. 1983. On what some concepts might not be. *Cognition* 13: 263–308.

Baltas, A., Gavroglu, K., & Kindi, V. 1997. A physicist who became a historian for philosophical purposes: A discussion between Thomas S. Kuhn and Aristides Baltas, Kostas Gavroglu, and Vassiliki Kindi. *Neusis* 6: 145–200. Reprinted as: A Discussion with Thomas S. Kuhn. In J. Conant & J. Haugeland (eds.): *The Road Since Structure*, pp. 253–323. Chicago: University of Chicago Press 2000.

Barker, P. 1990. Copernicus, the orbs and the equant. *Synthese* 83: 317–323.

Barker, P. 1998. Kuhn and the sociological revolution. *Configurations* 6: 21–32.

Barker, P. 1999. Copernicus and the critics of Ptolemy. *Journal for the History of Astronomy* 30: 343–358.

Barker, P. 2000. The role of religion in the Lutheran response to Copernicus. In M. J. Osler (ed.): *Rethinking the Scientific Revolution*, pp. 59–88. Cambridge: Cambridge University Press.

Barker, P. 2001. Incommensurability and conceptual change during the Copernican revolution. In H. Sankey & P. Hoyningen-Huene (eds.): *Incommensurability and Related Matters*, pp. 241–273. Boston Studies in the Philosophy of Science. Boston: Kluwer.

Barker, P. 2002. Constructing Copernicus. *Perspectives on Science* 10: 208–227.

Barker, P., Chen, X., & Andersen, H. 2003. Kuhn on concepts and categorization. In T. Nickles (ed.): *Thomas Kuhn*, pp. 212–245. Cambridge: Cambridge University Press.

Barker, P., & Goldstein, B. R. 1994. Distance and velocity in Kepler's astronomy. *Annals of Science* 51: 59–73.

Barker, P., & Goldstein, B. R. 1998. Realism and instrumentalism in sixteenth century astronomy: A reappraisal. *Perspectives on Science* 6: 232–258.

Barnes, B. 1974. *Scientific Knowledge and Sociological Theory*. London: Routledge & Kegan Paul.

Barnes, B.1982. *Thomas Kuhn and Social Science*. London: Macmillan.

Barsalou, L. W. 1982. Ad hoc categories. *Memory and Cognition* 11: 211–227.

Barsalou, L. W. 1985. Ideals, central tendency, and frequency of instantiation as determinants of graded structure in categories. *Journal of Experimental Psychology: Learning, Memory, and Cognition* 11: 629–654.

Barsalou, L. W. 1987. The instability of graded structure: Implications for the nature of concepts. In U. Neisser (ed.): *Concepts and Conceptual Development: Ecological and Intellectual Factors in Categorization*, pp. 101–140. Cambridge: Cambridge University Press.

Barsalou, L. W. 1988. The concept and organization of autobiographical memories. In U. Neisser & E. Winograd (eds.): *Remembering Reconsidered: Ecological and Traditional Approaches to the Study of Memory*, pp. 193–229. Cambridge: Cambridge University Press.

Barsalou, L. W. 1989. Intraconcept similarity and its implications for interconcept similarity. In S. Vosniadou & A. Ortony (eds.): *Similarity and Analogical Reasoning*, pp. 76–121. Cambridge: Cambridge University Press.

Barsalou, L. W. 1990. On the indistinguishability of exemplar memory and abstraction in category representation. In T. Srull & R. Wyer (eds.): *Advances in Social Cognition*, Vol. 3, pp. 61–88. Hillsdale, N.J.: Erlbaum.

Barsalou, L. W. 1991. Deriving categories to achieve goals. In G. H. Bower (ed.): *The Psychology of Learning and Motivation: Advances in Research and Theory*, Vol. 27, pp. 1–64. New York: Academic Press.

Barsalou, L. W. 1992a. *Cognitive Psychology: An Overview for Cognitive Scientists*. Hillsdale, N.J.: Erlbaum.

Barsalou, L. W. 1992b. Frames, concepts, and conceptual fields. In A. Lehrer & E. Kittay (eds.): *Frames, Fields and Contrasts: New Essays in Semantical and Lexical Organization*, pp. 21–74. Hillsdale, N.J.: Erlbaum.

Barsalou, L. W. 1993. Flexibility, structure and linguistic vagary in concepts: Manifestations of a compositional system of perceptual symbols. In A. F. Collins, S. E. Gathercole, M. A. Conway, & P. E. Morris (eds.): *Theories of Memory*, pp. 31–101. Hillsdale, N.J.: Erlbaum.

Barsalou, L. W. 1999. Perceptual symbol systems. *Behavioral and Brain Sciences* 22: 577–609.

Barsalou, L. W., & Billman, D. 1989. Systematicity and semantic ambiguity. In D. Gorfein (ed.): *Resolving Semantic Ambiguity*, pp. 146–203. New York: Springer.

Barsalou, L. W., & Hale, C. 1993. Components of conceptual representation: From feature-lists to recursive frames. In I. Mechelen, J. Hampton, R. Michalski, & P. Theuns (eds.): *Categories and Concepts: Theoretical Views and Inductive Data Analysis*, pp. 97–144. New York: Academic Press.

Barsalou, L. W., & Sewell, D. 1984. *Constructing Representations of Categories from Different Points of View. Emory Cognition Project Report No. 2.* Atlanta: Emory University.

Barsalou, L. W., & Sewell, D. 1985. Contrasting the representation of scripts and categories. *Journal of Memory and Language* 24: 646–665.

Barsalou, L. W., Solomon, K., & Wu, L. 1999. Perceptual simulation in conceptual tasks. In M. Hiraga, C. Sinha, & S. Wilcox (eds.): *Cultural, Typological, and Psychological Perspectives in Cognitive Linguistics: The Proceedings of the Fourth Conference of the International Cognitive Linguistics Association*, Vol. 3, pp. 209–228. Amsterdam: John Benjamins.

Bartlett, F. C. 1932. *Remembering.* London: Cambridge University Press.

Bechtel, W. 1988. *Philosophy of Mind: An Overview for Cognitive Science.* Hillsdale, N.J.: Erlbaum.

Bechtel, W., & Abrahamsen, A. 1991. *Connectionism and the Mind: An Introduction to Parallel Processing in Networks.* Oxford: Blackwell.

Becquerel, H. 1896a. Sur les radiations émises par phosphorescence. *Comptes Rendus* 122: 420–421.

Becquerel, H. 1896b. Sur les radiations invisibles émises par les corps phosphorescents. *Comptes Rendus* 122: 501–503.

Becquerel, H. 1896c. Sur quelques propriétés nouvelles des radiations invisibles émises par divers corps phosphorescents. *Comptes Rendus* 122: 559–564.

Becquerel, H. 1896d. Sur diverses propriétés des rayons uraniques. *Comptes Rendus* 123: 855–858.

Biagioli, M. 1990. The anthropology of incommensurability. *Studies in History and Philosophy of Science* 21: 183–209.

Biagioli, M. 1994. *Galileo Courtier.* Chicago: University of Chicago Press.

Black, E. 1977. *Manual of Neotropical Birds*, Vol. 1. Chicago: University of Chicago Press.

Bloor, D. 1976/1991. *Knowledge and Social Imagery.* London: Routledge & Kegan Paul. 2nd ed., Chicago: University of Chicago Press, 1991.

Bloor, D. 1983. *Wittgenstein: A Social Theory of Knowledge.* New York: Columbia University Press.

Bloor, D. 2002. *Wittgenstein, Rules and Institutions.* London: Routledge.

Bohr, N. 1936. Neutron capture and nuclear constitution. *Nature* 137: 344–348.

Bohr, N., & Wheeler, J. A. 1939. The mechanism of nuclear fission. *Physical Review* 56: 426–450.

di Bono, M. 1995. Copernicus, Amico, Fracastoro and Tusi's device: Observations on the use and transmission of a model. *Journal for the History of Astronomy* 26: 133–154.

Bourdieu, P. 2001. *Science de la science et réflexivité.* Paris: Raisons d'agir.

Bower, C., Black, J., & Turner, T. 1979. Scripts in memory for text. *Cognitive Psychology* 11: 177–220.

Boyd, R. 1979. Metaphor and theory change: What is "metaphor" a metaphor for? In A. Ortony (ed.): *Metaphor and Thought*, pp. 356–408. Cambridge: Cambridge University Press.

Brahe, T. 1588. *De mundi aetherei recentioribus phaenomenis.* Uraniburg:i.a.

Braybrooke, D., & Rosenberg, A. 1972. Getting the war news straight: The actual situation in philosophy of science. *American Political Science Review* 66: 818–826.

Brewer, W. 2000. Bartlett's concept of the schema and its impact on theories of knowledge representation in contemporary cognitive psychology. In A. Saito (ed.): *Bartlett, Culture and Cognition*, pp. 69–89. Hove, England: Psychology Press.

Brooks, L. R. 1987. Decentralized control of categorization: The role of prior processing episodes. In U. Neisser (ed.): *Concepts and Conceptual Development: Ecological and Intellectual Factors in Categorization*, pp. 141–174. Cambridge: Cambridge University Press.

Brown, N. R., Shevell, S. K., & Rips, L. J. 1987. Public memories and their personal context. In D. Rubin (ed.): *Autobiographical Memory*, pp. 137–158. Cambridge: Cambridge University Press.

Buchwald, J. 1992. Kinds and the wave theory of light. *Studies in History and Philosophy of Science* 23: 39–74.

Cantor, N., Smith, E. E., French, R. D., & Mezzich, J. 1980. Psychiatric diagnosis as prototype organization. *Journal of Abnormal Psychology* 89: 181–193.

Carey, S. 1985. *Conceptual Change in Childhood.* Cambridge, Mass.: The MIT Press.

Carey, S. 1991/1999. Knowledge acquisition: Enrichment or conceptual change? In E. Margolis & S. Laurence (eds.): *Concepts: Core Readings*, pp. 459–487. Cambridge, Mass.: MIT Press.

Carr, E. H. 1961. *What Is History?* London: Macmillan.

Chen, X. 1995. Taxonomic changes and the particle-wave debate in early nineteenth-century Britain. *Studies in History and Philosophy of Science* 26: 251–271.

Chen, X. 1997. Thomas Kuhn's latest notion of incommensurability. *Journal for General Philosophy of Science* 28: 257–273.

Chen, X. 2003. Why did John Herschel fail to understand polarization? The differences between object and event concepts. *Studies in History and Philosophy of Science* 34: 491–513.

Chen, X., Andersen, H., & Barker, P. 1998. Kuhn's theory of scientific revolutions and cognitive psychology. *Philosophical Psychology* 11: 5–28.

Chi, M., Feltovich, P., & Glaser, R. 1981. Categorization and representation of physics problems by experts and novices. *Cognitive Sciences* 5: 121–152.

Christianson, J. R. 1999. *On Tycho's Island.* Cambridge: Cambridge University Press.

Churchland, P. 1989. *A Neurocomputational Perspective: The Nature of Mind and the Structure of Science.* Cambridge, Mass.: MIT Press.

Clark, A. 1993. *Associative Engines: Connectionism, Concepts and Representational Change.* Cambridge, Mass.: MIT Press.

Collins, H. M. 1981. What is TRASP? The radical programme as a methodological imperative. *Philosophy of the Social Sciences* 11: 215–224.

Collins, H. M., & Pinch, T. 1993. *The Golem: What Everyone Should Know about Science,* Cambridge: Cambridge University Press.

Conklin, H. C. 1969. Lexicographical treatment of folk taxonomies. In S. A. Tyler (ed.): *Cognitive Anthropology,* pp. 41–49. New York: Holt, Rinehart & Winston.

Coulson, S. 2001. *Semantic Leaps: Frame Shifting and Conceptual Blending in Meaning Construction.* Cambridge: Cambridge University Press.

Covey, S. R. 1990. *The 7 Habits of Highly Effective People.* New York: Simon & Schuster.

Curd, M., & Cover, J. A. 1998. *Philosophy of Science: The Central Issues.* New York: W.W. Norton.

Curie, I., & Joliot, F. 1934. Un nouveau type de radioactivité. *Comptes Rendus* 198: 254–256.

Curie, M. 1898. Rayons émis par les composés de l'uranium et du thorium. *Comptes Rendus* 126: 1101–1103.

Curie, P., & Curie, M. 1898. Sur une substance nouvelle radio-active, contenue dans la pechblende. *Comptes Rendus* 127: 175–178.

Curie, P., Curie, M., & Bémont, G. 1898. Sur une nouvelle substance fortement radio-active, contenue dans la pechblende. *Comptes Rendus* 127: 1215–1217.

Darden, L. 1992. Strategies for anomaly resolution. In R. Giere (ed.): *Cognitive Models of Science,* pp. 251–273. Minnesota Studies in the Philosophy of Science, Vol. XV. Minneapolis: University of Minnesota Press.

Darden, L. 1998. Exemplars, abstractions, and anomalies: Representation and theory change in Mendelian and molecular genetics. In G. Wolters, J. G. Lennox, & P. McLaughlin (eds.): *Concepts, Theories and Rationality in the Biological Sciences: The Second Pittsburgh-Konstanz Colloquium in the Philosophy of Science,* pp. 137–158. Konstanz/Pittsburgh: Universitätsverlag Konstanz/ University of Pittsburgh Press.

Donahue, W. H. 1988. Kepler's fabricated figures: Covering up the mess in the *New Astronomy. Journal for the History of Astronomy* 19: 217–237.

Donahue, W. H. 1992. *Johannes Kepler – New Astronomy*. Cambridge: Cambridge University Press.

Drake, S. 1990. *Discoveries and Opinions of Galileo*. New York: Anchor Books.

Ekman, P., Friesen, W. V., & Ellsworth, P. 1972. *Emotion in the Human Face*. Elmsford, N.Y.: Pergamon.

Erreich, A., & Valian, V. 1979. Children's internal organization of locative categories. *Child Development* 50: 1070–1077.

Evans, J. 1998. *The History and Practice of Ancient Astronomy*. Oxford: Oxford University Press.

Evans, R. J. 1999. *In Defense of History*. New York: W. W. Norton.

Feather, N., & Bretcher, E. 1939. Atomic numbers of the so-called transuranic elements. *Nature* 143: 516.

Fermi, E. 1934. Possible production of elements of atomic number higher than 92. *Nature* 133: 898–899.

Fermi, E. 1939. Artificial radioactivity produced by neutron bombardment. Nobel Lecture. Reprinted in E. Fermi, *Collected Papers*, Vol. 1: *Italy 1921–1938*, pp. 1037–1043. Chicago: University of Chicago Press, 1962.

Flügge, S. 1939. Kann der Energiegehalt der Atomkerne technisch nutzbar gemacht werden. *Die Naturwissenschaften* 27: 402–410.

Gadow, H. 1892. On the classification of birds. *Proceedings of the General Meeting for Scientific Business of the Zoological Society of London 1892*, pp. 229–256.

Galilei, Galileo 1610/1989. *Sidereus nuncius, or the Sidereal Messenger*. Translated with introduction, conclusion, and notes by Albert Van Helden. Chicago: University of Chicago Press.

Gamow, G. 1929a. Über die Struktur des Atomkerns. *Physikalische Zeitschrift* 30: 717–720.

Gamow, G. 1929b. Zur Quantentheorie der Atomzertrümmerung. *Zeitschrift der Physik* 52: 510–515.

Gamow, G. 1931. *Constitution of Atomic Nuclei and Radioactivity*. Oxford: Clarendon.

Gentner, D. 1988. Metaphor as structure mapping: The relational shift. *Child Development* 59: 47–59.

Giere, R. 1988. *Explaining Science: A Cognitive Approach*. Chicago: University of Chicago Press.

Giere, R. (ed.) 1992. *Cognitive Models of Science*. Minneapolis: University of Minnesota Press.

Giere, R., 1994. The cognitive structure of scientific theories. *Philosophy of Science* 61: 276–296.

Gingerich, O. 1975. "Crisis" versus aesthetic in the Copernican Revolution. In A. Beer (ed.): *Vistas in Astronomy*, Vol. 17, pp. 85–94. Oxford: Pergamon Press.

Gingerich, O. 1993. Erasmus Reinhold and the dissemination of the Copernican theory. In O. Gingerich: *The Eye of Heaven*. New York: American Institute of Physics.

Goldstein, B. R. 1991. The blasphemy of Alfonso X: History or myth? In Peter Barker & Roger Ariew (eds.): *Revolution and Continuity: Essays in the*

History and Philosophy of Early Modern Science, pp. 143–153. Washington, D.C.: Catholic University of America Press.

Goldstein, B. R., & Barker, P. 1995. The role of Rothmann in the dissolution of the celestial spheres. *British Journal for the History of Science* 28: 385–403.

Goldstone, R. L., Medin, D. L., & Gentner, D. 1991. Relational similarity and the nonindependence of features in similarity judgments. *Cognitive Psychology* 23: 222–262.

Golinski, J. 1990. The theory of practice and the practice of theory: Sociological approaches in the history of science. *Isis* 81: 492–505.

Gooding, D. 1990. *Experiment and the Making of Meaning*. Dordrecht: Kluwer.

Gopnik, A., & Meltzoff, A. 1997. *Words, Thoughts and Theories*. Cambridge, Mass.: The MIT Press.

Hacking, I. 1993. Working in a new world: The taxonomic solution. In P. Horwich (ed.): *World Changes: Thomas Kuhn and the Nature of Science*, pp. 275–310. Cambridge, Mass.: The MIT Press.

Hahn, D. (ed.) 1975. *Otto Hahn: Erlebnisse und Erkenntnisse*. Düsseldorf: Econ Verlag.

Hahn, O. 1946. Von den natürlichen Umwandlungen des Urans zu seiner künstlichen Zerspaltung, Nobel-Vortrag am 13. Dez. 1946. In O. Hahn: *Mein Leben*, pp. 247–267. Bruckmann: München 1968.

Hahn, O., & Strassmann, F. 1938. Über die Entstehung von Radiumisotopen aus Uran durch Bestrahlen mit schnellen und verlangsammten Neutronen. *Die Naturwissenschaften* 26: 755–756.

von Halban, H., Joliot, F., & Kowarski, L. 1939a. Liberation of neutrons in the nuclear explosion of uranium. *Nature* 143: 70–71.

von Halban, H., Joliot, F., & Kowarski, L. 1939b. Number of neutrons liberated in the nuclear fission of uranium. *Nature* 143: 680.

Hamel, J. 1998. *Die astronomische Forschungen in Kassel unter Wilhelm IV.* Thun and Frankfurt am Main: Harri Deutsch.

Heider, E. R. 1972. Universals in color naming and memory. *Journal of Experimental Psychology* 93: 10–20.

van Helden, A. 1985. *Measuring the Universe*. Chicago: University of Chicago Press.

Henry, C. 1896. Augmentation du rendement photographique des rayons Röntgen par le sulfure de zinc phosphorescent. *Comptes Rendus* 122: 312–314.

Homa, D., & Vosburgh, R. 1976. Category breadth and the abstraction of prototypical information. *Journal of Experimental Psychology – Human Learning and Memory* 2: 322–330.

Hoyningen-Huene, P. 1993. *Reconstructing Scientific Revolutions: Thomas S. Kuhn's Philosophy of Science*. Chicago: University of Chicago Press.

Hoyningen-Huene, P., & Sankey, H. 2001. *Incommensurability and Related Matters*. Dordrecht: Kluwer.

Kay, P. 1971. Taxonomy and semantic contrast. *Language* 447: 866–887.

Keil, F. 1989. *Concepts, Kinds and Cognitive Development.* Cambridge, Mass.: The MIT Press.

Kendler, T., & Kendler, H. 1970. An ontogeny of optical shift behavior. *Child Development* 41: 1–27.

Kepler, J. 1596. *Mysterium Cosmographicum.* Tübingen: G. Gruppenbachius.

Kepler, J. 1609. *Astronomia Nova.* Heidelberg: G. Voegelinus. Translated by W. H. Donahue, as: *Johannes Kepler: New Astronomy.* Cambridge: Cambridge University Press, 1992.

Kepler, J. 1618–1622. *Epitome Astronomiae Copernicanae.* Linz: Plancus, & Frankfurt: Tampachius.

Krafft, F. 1981. *Im Schatten der Sensation: Leben und Wirken von Fritz Strassmann.* Weinhein: Verlag Chemie.

Kragh, H. 1999. *Quantum Generations: A History of Physics in the Twentieth Century.* Princeton, N.J.: Princeton University Press.

Kuhn, T. S. 1957. *The Copernican Revolution: Planetary Astronomy in the Development of Western Thought.* Cambridge, Mass.: Harvard University Press.

Kuhn, T. S. 1959. The essential tension: Tradition and innovation in scientific research. In C. W. Taylor & F. Barron (eds.): *Scientific Creativity: Its Recognition and Development,* pp. 341–354. New York: John Wiley. Reprinted in Kuhn (1977), pp. 225–259.

Kuhn, T. S. 1962. *The Structure of Scientific Revolutions.* Chicago: University of Chicago Press.

Kuhn, T. S. 1970a. *The Structure of Scientific Revolutions.* 2nd ed. Chicago: University of Chicago Press.

Kuhn, T. S. 1970b. Reflections on my critics. In I. Lakatos & A. Musgrave (eds.): *Criticism and the Growth of Knowledge,* pp. 231–278. Cambridge: Cambridge University Press.

Kuhn, T. S. 1974. Second thoughts on paradigms. In F. Suppe (ed.): *The Structure of Scientific Theories,* pp. 459–482. Urbana: University of Illinois Press. Reprinted in Kuhn (1977), pp. 293–319.

Kuhn, T. S. 1976. Theory-change as structure-change: Comments on the Sneed formalism. *Erkenntnis* 10: 179–199.

Kuhn, T. S. 1977. *The Essential Tension: Selected Studies in Scientific Tradition and Change.* Chicago: University of Chicago Press.

Kuhn, T. S. 1979. Metaphor in science. In A. Ortony (ed.): *Metaphor and Thought,* pp. 533–542. Cambridge: Cambridge University Press, 1979. Reprinted in Kuhn (2000), pp. 196–207.

Kuhn, T. S. 1983a. Commensurability, comparability, communicability. *PSA 1982* 2: 669–688.

Kuhn, T. S. 1983b. Response to commentaries. In *PSA 1982* 2: 712–716. Reprinted in Kuhn (2000), pp. 53–57.

Kuhn, T. S. 1989. Possible worlds in history of science. In S. Allén (ed.): *Possible Worlds in Humanities, Arts and Sciences: Proceedings of Nobel Symposium 65,* pp. 9–32. Berlin: de Gruyter. Reprinted in Kuhn (2000), pp. 58–89.

Kuhn, T. S. 1990. Dubbing and redubbing: The vulnerability of rigid designation. In C. W. Savage (ed.): *Scientific Theories,* pp. 298–318. Minnesota

Studies in the Philosophy of Science, Vol. XIV. Minneapolis: University of Minnesota Press.

Kuhn, T. S. 1991. The road since *Structure*. *PSA 1990* 2: 3–13.

Kuhn, T. S. 1992. The trouble with the historical philosophy of science. Robert and Maureen Rothschild Distinguished Lecture, 19 Nov. 1991. Cambridge, Mass.: Department of the History of Science, Harvard University. Reprinted in Kuhn (2000), pp. 105–120.

Kuhn, T. S. 1993. Afterwords. In P. Horwich (ed.): *World Changes*, pp. 311–341. Cambridge, Mass.: The MIT Press.

Kuhn, T. S. 2000. *The Road since Structure*. J. Conant & J. Haugeland (eds.). Chicago: University of Chicago Press.

Kuhn, T. S., Shapere, D., Bromberger, S., Suppes, P., Putnam, H., & Achinstein, P. 1974. Discussion. In F. Suppe (ed.): *The Structure of Scientific Theories*, pp. 500–517. Urbana: University of Illinois Press.

Lakatos, I. 1978. *The Methodology of Scientific Research Programmes*. Cambridge: Cambridge University Press.

Lakatos, I., & Musgrave, A. 1970. *Criticism and the Growth of Knowledge*. Cambridge: Cambridge University Press.

Lakoff, G. 1987. *Women, Fire and Dangerous Things: What Categories Reveal about the Mind*. Chicago: University of Chicago Press.

Latour, B. 1986. *Laboratory Life: The Construction of Scientific Facts*. Princeton N.J.: Princeton University Press.

Lattis, J. 1994. *Between Copernicus and Galileo: Christopher Clavius and the Collapse of Ptolemaic Astronomy*. Chicago: University of Chicago Press.

Laudan, L. 1977. *Progress and Its Problems: Towards a Theory of Scientific Growth*. London: Routledge & Keagan Paul.

Van Loocke, P. (ed.) 1999. *The Nature of Concepts: Evolution, Structure and Representation*. London: Routledge.

Maestlin, M. 1578. *Observatio et demonstratio cometae aetherei . . . anno 1577 et 1588*. Tübingen: Gruppenbach.

Magnani, L., Nersessian, N. J., & Thagard, P. 1999. *Model-Based Reasoning in Scientific Discovery*. New York: Plenum.

Malt, B. C., & Smith, E. E. 1984. Correlated properties in natural categories. *Journal of Verbal Learning and Verbal Behavior* 23: 250–269.

Margolis, E., & Laurence, S. (eds.) 1999. *Concepts: Core Readings*. Cambridge, Mass.: The MIT Press.

Margolis, H. 2002. *It Started with Copernicus: How Turning the World Inside Out Led to the Scientific Revolution*. Chicago: University of Chicago Press.

McMillan, E. 1939. Radioactive recoil from uranium activated by neutrons. *Physical Review* 55: 510.

Medin, D. L. 1989. Concepts and conceptual structure. *American Psychologist* 44: 1469–1481.

Medin, D., Altom, M., Edelson, S., & Freko, D. 1982. Correlated symptoms and simulated medical classification. *Journal of Experimental Psychology: Learning, Memory, and Cognition* 8: 37–50.

Medin, D. L., & Schaffer, M. M. 1978. Context theory of classification learning. *Psychological Review* 86: 207–238.

Meitner, L., & Delbrück, M. 1935. *Der Aufbau der Atomkerne. Natürliche und künstliche Kernumwandlungen.* Berlin: Springer.

Meitner, L., & Frisch, O. 1939. Disintegration of uranium by neutrons: A new type of nuclear reaction. *Nature* 143: 239–240.

Meitner, L., & Hahn, O. 1936. Neue Umwandlungsprozess bei Bestrahlung des Urans mit Neutronen. *Naturwissenschaften* 24: 158–159.

Melanchthon, P. 1549/1846. *Initia Doctrinae Physicae.* In C. G. Bretschneider (ed.): *Corpus Reformatorum*, Vol. 13; pp. 1179–1412. Halis Saxonum: Schwetschke et Filius. Reprint, New York: Johnson Reprint, 1964.

Mervis, C. B., & Pani, J. R. 1980. Acquisition of basic object categories. *Cognitive Psychology* 12: 496–522.

Mervis, C. B., & Rosch, E. 1981. Categorization of natural objects. *Annual Review of Psychology* 32: 89–115.

Minsky, M. 1975. A framework for representing knowledge. In P. Winston (ed.): *The Psychology of Computer Vision*, pp. 211–277. New York: McGraw-Hill.

Murphy, G. L., & Medin, D. L. 1985. The role of theories in conceptual coherence. *Psychological Review* 92: 289–316.

Murphy, G., & Medin, D. (1985/1999). The role of theories in conceptual coherence. In E. Margolis & S. Laurence (eds.): *Concepts: Core Readings*, pp. 425–458. Cambridge, Mass.: The MIT Press.

Murphy, G., & Smith, E. 1982. Basic-level superiority in picture categorization. *Journal of Verbal Learning and Verbal Behavior* 21: 1–20.

Nersessian, N. J. 1984. *Faraday to Einstein: Constructing Meaning in Scientific Theories.* Dordrecht: Martinus Nijhoff.

Nersessian, N. J. 1987. A cognitive-historical approach to meaning in scientific theories. In N. Nersessian (ed.): *The Process of Science*, pp. 161–177. Dordrecht, Kluwer.

Nersessian, N. J. 1989. Conceptual change in science and in science education. *Synthese* 80: 163–183.

Nersessian, N. J. 1992a. How do scientists think? Capturing the dynamics of conceptual change in science. In R. N. Giere (ed.): *Cognitive Models of Science*, pp. 3–45. Minneapolis: University of Minnesota Press.

Nersessian, N. J. 1992b. In the theoretician's laboratory: Thought experimenting as mental modeling. *PSA* 2: 291–301.

Nersessian, N. J. 1995. Opening the black box: Cognitive science and the history of science. In A. Thackray (ed.): *Constructing Knowledge in the History of Science. Osiris* 10: 194–214.

Nersessian, N. J. 1998. Conceptual change. In W. Bechtel & G. Graham (eds.): *A Companion to Cognitive Science*, pp. 155–166. Oxford: Blackwell.

Nersessian, N. J. 1999. Model-based reasoning in conceptual change. In L. Magnani, N. J. Nersessian, & P. Thagard (eds.): *Model-Based Reasoning in Scientific Discovery*, pp. 5–22. New York: Kluwer.

Nersessian, N. J. 2001. Maxwell and "the method of physical analogy": Model-based reasoning, generic abstraction, and conceptual change.

In D. Malament (ed.): *Reading Philosophy of Nature: Essays in the History and Philosophy of Science and Mathematics to Honor Howard Stein on his 70th Birthday*, pp. 129–166. LaSalle, IL: Open Court.

Nersessian, N. J. 2003. Kuhn, conceptual change and cognitive science. In T. Nickles (ed.): *Thomas Kuhn*, pp. 178–211. Cambridge: Cambridge University Press.

Nersessian, N. J., & Andersen, H. 1998. Conceptual change and incommensurability: A cognitive-historical view. *Danish Yearbook of Philosophy* 32: 111–151.

Nersessian, N. J., & Magnani, L. 2002. *Model-Based Reasoning: Science, Technology, and Values*. New York: Kluwer.

Neugebauer, O. 1968. On the planetary theory of Copernicus. *Vistas in Astronomy* 10: 89–103.

Newton, A. 1893. *A Dictionary of Birds*. London: Adam and Charles Black.

Nickles, T. 2003. Normal science: From logic to case-based and model-based reasoning. In T. Nickles (ed.): *Thomas Kuhn*, pp. 142–177. Cambridge: Cambridge University Press.

Niewenglowski, G. H. 1896. Sur la propriété qu'ont les radiations émises par les corps phosphorescents, de traverser certains corps opaques à la lumière solaire, et sur les expériences de M.G. le Bon, sur la lumière noire. *Comptes Rendus* 122: 385–386.

Noddack, I. 1934a. Das Periodische System der Elemente und seine Lücken. *Angewandte Chemie* 47: 301–305.

Noddack, I. 1934b. Über das Element 93. *Angewandte Chemie* 47: 653–655.

Nye, M.-J. 1980. N-rays: An episode in the history and psychology of science. *Historical Studies in the Physical Sciences* 11: 125–156.

Pais, A. 1977. Radioactivity's two early puzzles. *Reviews of Modern Physics* 49: 925–938.

Pais, A. 1986. *Inward Bound: Of Matter and Force in the Physical World*. Oxford: Oxford University Press.

Pedersen, O. 1993. *Early Physics and Astronomy*. Cambridge: Cambridge University Press.

Pickering, A. 1984. *Constructing Quarks: A Sociological History of Particle Physics*. Chicago: University of Chicago Press.

Pinch, T. 1986. *Confronting Nature: The Sociology of Solar-Neutrino Detection*. Dordrecht, Kluwer.

Poincaré, H. 1896. Les rayons cathodiques et les rayons de Röntgen. *Revue générale des Sciences pures et appliquées* 7: 52–59.

Putnam, H. 1975. The meaning of "meaning." In K. Gundersen (ed.): *Language, Mind and Knowledge*, pp. 131–193. Minnesota Studies in the Philosophy of Science, Vol. VII. Minneapolis: University of Minnesota Press.

Ragep, F. J. 1993. *Nasir al-Din al Tusi's Memoir on Astronomy*. New York: Springer.

Ray, J. 1678. *The Ornithology of Francis Willughby*. London: John Martyn.

Reinhold, E. 1542. *Theoricae novae planetarum Georgii Purbacchii (sic) Germani ab Erasmo Reinholdo Salveldensi. Inserta item methodica tractio de illuminiatione Lunae. Typus Eclipsis solis futurae Anno 1544*. Wittenberg: Luft.

Rey, G. 1985. Concepts and conceptions: A reply to Smith, Medin and Rips. *Cognition* 19: 297–303.

Rheticus, G. J. 1540/1979. *Narratio prima.* H. Hugonnard-Roche et al. (eds. and trans.). Wroclaw: Ossolineum, Studia Copernicana 20.

Rhodes, R. 1986. *The Making of the Atomic Bomb.* New York: Simon & Schuster.

Rips, L. J. 1975. Inductive judgements about natural categories. *Journal of Verbal Learning and Verbal Behavior* 14: 665–681.

Röntgen, W. K. 1896. *Eine neue Art von Strahlen.* Würzburg: Stahel'schen K. Hof- und Universitätsbuch- und Kunsthandlung.

Rosch, E. 1973a. Natural categories. *Cognitive Psychology* 4: 328–350.

Rosch, E. 1973b. On the internal structure of perceptual and semantic categories. In T. E. Moore (ed.): *Cognitive Development and the Acquisition of Language,* pp. 111–144. New York: Academic Press.

Rosch, E. 1978. Principles of categorization. In E. Rosch & B. B. Lloyd (eds.): *Cognition and Categorization,* pp. 27–48. Hillsdale, N.J.: Erlbaum.

Rosch, E. 1987. Wittgenstein and categorization research in cognitive psychology. In M. Chapman & R. A. Dixon (eds.): *Meaning and the Growth of Understanding: Wittgenstein's Significance for Developmental Psychology,* pp. 151–166. Berlin: Springer.

Rosch, E., & Mervis, C. B. 1975. Family resemblances: Studies in the internal structures of categories. *Cognitive Psychology* 7: 573–605.

Rosch, E., Mervis, C., Gray, W., Johnson, D., & Boyes-Braem, P. 1976. Basic objects in natural categories. *Cognitive Psychology* 8: 382–439.

Ross, B., Perkins, S., & Tenpenny, P. 1990. Reminding-based category learning. *Cognitive Psychology* 22: 460–492.

Rudwick, M. J. S. 1985. *The Great Devonian Controversy: The Shaping of Scientific Knowledge among Gentlemanly Specialists.* Chicago: University of Chicago Press.

Rutherford, E. 1899. Uranium radiation and the electrical conduction produced by it. *Philosophical Magazine* 47: 109–163. Reprinted in J. Chadwick (ed.): *The Collected Papers of Lord Rutherford of Nelson,* pp. 169–215. London: George Allen & Unwin, 1962.

Rutherford, E. 1903. The magnetic and electric deviation of the easily absorbed rays from Radium. *Philosophical Magazine* 5: 177–187. Reprinted in J. Chadwick (ed.): *The Collected Papers of Lord Rutherford of Nelson,* pp. 549–557. London: George Allen & Unwin, 1962.

Sankey, H. 1994. *The Incommensurability Thesis.* Aldershot, England: Avesbury.

Schank, R. C. 1975. *Conceptual Information Processing.* Amsterdam: North Holland.

Schank, R., & Abelson, R. 1977. *Scripts, Plans, Goals, and Understanding.* Hillsdale, N.J.: Erlbaum.

Scheffler, I. 1967. *Science and Subjectivity.* Indianapolis: Bobbs-Merrill.

Seaborg, G. 1989. Nuclear fission and transuranic elements – 50 years ago. *Journal of Chemical Education* 66: 379–384.

Shapere, D. 1964. The structure of scientific revolutions. *Philosophical Review* 73: 383–394.

Shapere, D. 1982. Reason, reference, and the quest for knowledge. *Philosophy of Science* 49: 1–23.

Shapere, D. 1989. Evolution and continuity in scientific change. *Philosophy of Science* 56: 419–437.

Shapin, S. 1975. Phrenological knowledge and the social structure of early 19th-century Edinburgh. *Annals of Science* 32: 219–243.

Shapin, S. 1982. History of science and its sociological reconstructions. *History of Science* 20: 157–211.

Shapin, S., & Schaffer, S. 1984. *Leviathan and the Air-Pump: Hobbes, Boyle, and the Experimental Life.* Princeton: Princeton University Press.

Shepp, B. 1978. From perceived similarity to dimensional structure: A new hypothesis about perspective development. In E. Rosch & B. Lloyd (eds.): *Cognition and Categorization*, pp. 135–167. Hillsdale, NJ: Erlbaum.

Sibley, C., & Ahlquist, J. 1990. *Phylogeny and Classification of Birds: A Study in Molecular Evolution.* New Haven: Yale University Press.

Sloman, S. A., Love, B. C., & Ahn, W. K. 1998. Feature centrality and conceptual coherence. *Cognitive Science* 22: 189–227.

Smith, B. H. 1997. *Belief and Resistance: Dynamics of Contemporary Intellectual Controversy.* Cambridge, Mass.: Harvard University Press.

Smith, E., Osherson, D., Rips, L., & Keane, M. 1988. Combining prototypes: A selective modification model. *Cognitive Sciences* 12: 485–527.

Smith, J. D., & Kemler-Nelson, D. 1984. Overall similarity in adults' classification: The child in all of us. *Journal of Experimental Psychology: General* 113: 137–159.

Solomon, M. 2001. *Social Empiricism.* Cambridge, Mass.: The MIT Press.

Stein, N. 1992. What's in a story: Interpreting the interpretations of story grammars. *Discourse Processes* 5: 319–335.

Stevenson, B. 1994. *Kepler's Physical Astronomy.* Princeton, N.J.: Princeton University Press.

Stuewer, R. 1994. The origin of the liquid-drop model and the interpretation of nuclear fission. *Perspectives on Science* 2: 76–129.

Sundevall, C. 1889. *Sundevall's Tentamen.* London: Porter.

Swerdlow, N. 1976. Pseudodoxia Copernicana. *Archives Internationales d'Historie des Sciences* 26: 105–158.

Taves, R. 1998. "Frank and Ernest" cartoon published 2-26-1998. NEA Inc.

Thagard, P. 1992. *Conceptual Revolutions.* Princeton, N.J.: Princeton University Press.

Thoren, V. E. 1990. *The Lord of Uraniborg: A Biography of Tycho Brahe.* Cambridge: Cambridge University Press.

Treumann, R. A. 1991. A post-fission perspective of the discovery of nuclear fission. *Journal for the General Philosophy of Science* 22: 143–153.

Tversky, A. 1977. Features of similarity. *Psychological Review* 84: 327–352.

Tversky, B., & Hemenway, K. 1984. Objects, parts, and categories. *Journal of Experimental Psychology: General* 113: 169–193.

Voelkel, J. R. 2001. *The composition of Kepler's Astronomia nova.* Princeton, N.J.: Princeton University Press.

Waismann, F. 1965. *The Principles of Linguistic Philosophy.* R. Harré (ed.). London: Macmillan.

Way, E. C. 1997. Connectionism and conceptual structure. *American Behavioral Scientist* 40: 729–753.

Weart, S. 1983. The discovery of physics and a nuclear physics paradigm. In W. R. Shea (ed.): *Otto Hahn and the Rise of Nuclear Physics,* pp. 91–133. Dordrecht: Kluwer.

von Weizsäcker, F. 1937. *Die Atomkerne.* Berlin: Springer.

Westman, R. S. 1975. The Melanchthon Circle, Rheticus, and the Wittenberg Interpretation of the Copernican Theory. *Isis* 66: 165–193.

Westman, R. 1994. Two cultures or one? A second look at Kuhn's *The Copernican Revolution. Isis* 85: 79–155.

Williams, T. M., Freyer M. L., & Aiken, L. S. 1977. Development of visual pattern classification in preschool children: Prototypes and distinctive features. *Developmental Psychology* 13: 577–584.

Wisniewski, E., & Medin, D. 1991. Harpoons and long sticks: The interaction of theory and similarity in rule induction. In D. Fischer, M. Pazzani, & P. Langley (eds.): *Concept Formation: Knowledge and Experience in Unsupervised Learning,* pp. 237–278. San Mateo, Calif.: Morgan Kaufmann.

Wittgenstein, L. W. 1953. *Philosophical Investigations.* G. E. M. Anscombe (trans.). Oxford: Blackwell.

Index

Page numbers followed by "*f*" denote figures.

195

Printed in the United States
By Bookmasters